THE BIOCHEMICAL APPROACH TO LIFE

THE
BIOCHEMICAL APPROACH
TO LIFE

BY

F. R. JEVONS

Lecturer in Biological Chemistry, University of Manchester
formerly University Demonstrator in Biochemistry
and Fellow of King's College, Cambridge

WITH A FOREWORD BY
F. SANGER, F.R.S.

London
GEORGE ALLEN & UNWIN LTD
RUSKIN HOUSE   MUSEUM STREET

FIRST PUBLISHED IN 1964

This book is copyright under the Berne Convention. Apart from any fair dealing for the purposes of private study, research, criticism or review, as permitted under the Copyright Act, 1956, no portion may be reproduced by any process without written permission. Enquiries should be addressed to the publisher.

© *George Allen & Unwin Ltd, 1964*

TO DITA

PRINTED IN NORTHERN IRELAND
in 10 point Times Roman type
AT THE UNIVERSITIES PRESS, BELFAST

Foreword

WHAT is biochemistry? A few decades ago this was not a question that many people would ask, but today nearly everyone has heard about this young science and there must be very many who want to know more about it. It is at present particularly exciting since it is developing at an extremely rapid rate and new and interesting findings are constantly coming to light. Biochemists are seeking the answers to many fundamental questions that we all ask ourselves as soon as we start wondering about things. What are we made of? How do our bodies work? What is life? Already the answers to some of these questions are beginning to emerge. Much work has been done and the biochemical approach has been shown to be a highly successful one; many new lines of research are opening up and even more work remains for the future. Most of the foundations of the science have been laid but its full development and its application to practical affairs in medicine and industry are largely for the future.

Since biochemistry is considered to be a borderline science embodying both biology and chemistry, it is rarely given as prominent a place as it deserves in any school curriculum or even in the universities, so that many science students are never introduced to the subject and its fascinations until they have already decided in which fields to specialize. This is one of the reasons why this book is so extremely welcome. It is clearly and easily written and should be comprehensible to anyone with an elementary knowledge of science, yet it gives a lucid and stimulating introduction to biochemistry.

Dr. Jevons has himself made important contributions to this science in the laboratory and much of his enthusiasm for the subject appears in this book, which helps to make it extremely good reading not only to the novice but also to more experienced scientists, who will certainly find something new in it, especially where the history of biochemistry is dealt with. This is a recent interest of Dr. Jevons and the strongly historical flavour of this book increases its readability and helps to put the whole of biochemistry into its proper perspective.

Indeed, this is the sort of book that I would like to have written myself and I regard it as an honour to be associated with it and to make my contribution in this easy manner of writing a brief foreword.

F. SANGER

31*st* January, 1963

Preface

BIOCHEMISTRY is no longer a sideline of science but one of its main advancing frontiers. One need not be an expert to be aware of this. It is only necessary to look, for instance, at the large number of recent Nobel prizes given for work of predominantly biochemical nature; they include nearly half the awards made since the Second World War both for chemistry and for medicine. (Young hopefuls can thus put two promising irons in the fire by choosing biochemistry.)

The biochemical approach, then, is clearly proving fruitful. In this book, I have tried to give some insight into its nature. To many, biochemistry presents the forbidding image of a vastly complicated kind of chemistry. Here, it is presented rather as a very fundamental kind of biology. The emphasis is on biochemistry as a way of explaining the phenomena of life. To give conceptual coherence to the individual topics treated, I have tried to build up a rationale of the biochemical approach. From isolated molecules and events on the molecular scale, typified by proteins and single enzyme reactions, the discussion moves on to their collaboration and organization above the molecular level in subcellular particles. Later, attention is directed to the problems of finding out how molecular events underlie macroscopic phenomena, with special reference to the modes of action of vitamins, drugs and genetic factors.

A discussion of biochemistry can hardly rise above the plane of the trivial without using some of the language of chemistry. The technicalities have been kept to a minimum, however, and the chemical and biological background assumed is not above school level. My aim has been *la haute vulgarisation*—to provide food for thought for those who would like to apply their minds intelligently to the subject even though they may start with little specialized knowledge about it.

Parts of the manuscript have been read by Dr. G. R. Barker, Dr. W. Mays and Dr. C. H. Wynn of the University of Manchester, and by Dr. Sanger of Cambridge. I am grateful to them for their comments, which have helped me to clarify a number of points. The greatest debt of which I feel conscious is to all the students I have taught; any merit of exposition this book may have owes much to them. Those who asked questions helped me most, but even those who merely looked puzzled helped a little, by stimulating me to find more effective ways of making myself understood.

F. R. JEVONS

Contents

FOREWORD   page 7
PREFACE   9

I. BIOCHEMISTRY IN RELATION TO BIOLOGY AND CHEMISTRY   15

II. BIG MOLECULES—THE PROTEINS   18
   Large and labile—handle with care! Purifying proteins.
   Fibrinogen from blood, and ovalbumin from egg white.
   Protein "heat-stroke"—denaturation.
   Protein structure (i) The units in the chains—amino-acid analysis.
   A digression on chromatography—separations in columns and on paper.
   Protein structure (ii) Order in the chains—amino-acid sequence. Insulin.
   Protein structure (iii) Folding of the chains.
   Protein structure (iv) Cross-linking the chains—disulphide bridges. Keratin.
   Protein molecules as packets of electric charges.
   Handling big molecules (i) Running the electric gauntlet—ion exchange chromatography.
   Handling big molecules (ii) Moving in an electric field—electrophoresis.
   Handling big molecules (iii) Moving in a centrifugal field—ultracentrifugation.
   A biological event isolated—the clotting of fibrinogen. Enzymes as proteins.

III. LIFE'S BASIC DEVICE—SPECIFIC CATALYSIS BY ENZYMES. THE ANAEROBIC BREAKDOWN OF SUGAR   55
   A worldly motive—the thirst for alcohol.
   Chemical thesis—the overall result.
   Biological antithesis—yeast as a living organism.
   Biochemical synthesis—a multi-enzyme system.

CONTENTS

An unexpected participant—involvement of phosphate.
Getting along without oxygen—oxido-reduction aspect.
The biochemical unity of living matter—the similarity of yeast and muscle.
The role of enzymes in biological thought—nineteenth century "vitalism".

IV. ORGANIZATION AND EFFICIENCY—SUBCELLULAR PARTICLES AND BIOLOGICAL OXIDATIONS ... 72

Care in handling again—preparing subcellular particles.
Cells as more than bags of enzymes—enzymes in the particles.
Passing the buck—respiratory chains.
Into the neck of the funnel—formation of acetylcoenzyme A.
The last round of catabolism—the citrate cycle.

V. TWO APPROACHES TO BIOLOGICAL EXPLANATION—ANALOGY AND ANALYSIS ... 87

The biochemist's dilemma—to imitate or to take to pieces?
The limitations of analogy. Descartes' animal machine and its successors.
Analogies applied to biological oxidations.
The objections to analysis.
Meeting the objections (i) Intermediate levels to bridge the gap.
Meeting the objections (ii) Spectroscopic reconnaissance.
Meeting the objections (iii) Isotopic espionage.
Isotopic carbon in the study of photosynthesis.
A dated controversy—"mechanism" versus "vitalism". The organization hierarchy.

VI. THE HIGH ROAD AND THE LOW ROAD—VITAMINS AND COENZYMES ... 112

Two roads to a dramatic meeting.
The nutritional road—vitamins (i) Deficiency diseases.

CONTENTS

The nutritional road—vitamins (ii) Analysis of diet.
The enzymological road—coenzymes.
The two roads meet (i) Thiamine, TPP and beri-beri.
The two roads meet (ii) Nicotinate, NAD and pellagra.
The difficulties of biochemical prediction.

VII. DESIGNING DRUGS—THE DREAM OF A RATIONAL CHEMOTHERAPY   129

Old ideas behind old remedies.
The importance of selectivity.
From the mice that "died cured" to organic arsenicals.
From dyes to sulphonamides.
The importance of competition—an imperfect key to jam the lock.
A rationale at last? The antimetabolite rush.
Shifting the burden of design—antibiotics.

VIII. A COMMON CURRENCY FOR ENERGY TRANSACTIONS—ATP   143

The convenience of a common currency.
Where the value lies—high-energy bonds.
Pennies for the slot machines—exchanging glucose for ATP.
Getting small change (i) anaerobic phase.
Getting small change (ii) aerobic phase.
The loss on the deal—a question of efficiency again.
Spending the pennies (i) for chemical synthesis.
Spending the pennies (ii) for pumping.
Spending the pennies (iii) for light.
Spending the pennies (iv) for muscular contraction.

IX. TRANSMITTING INFORMATION—BIOCHEMICAL GENETICS   164

Miniature or message? The preformationist and modern views compared.
The bearer of the message—DNA.
Identifying the bearer—evidence for the genetic role of DNA.

CONTENTS

Copying the message—the duplication of DNA.
Decoding the message—the control of protein synthesis by nucleic acids.
The wrong messages—viruses.
Putting the instructions into effect—genes and enzymes.
The chicken and the egg—what is the basic minimum of life?

INDEX   183

I
Biochemistry in Relation to Biology and Chemistry

"BIOCHEMISTS"—so runs one elderly definition—"are people who talk about chemistry to biologists, about biology to chemists, and about women among themselves". It is hardly the sort of definition ever to have had much currency among biochemists themselves. To a biochemist to-day, who can afford to smile at the implied resentment of an upstart by two older-established sciences, it also seems as dated as a daguerreotype. It is not only that what used to be virtually an all-male preserve has been thoroughly infiltrated; more important, it is the idea of biochemistry as little more than a hotch-potch of borderline facts from two more basic sciences that quite fails to catch the right flavour. Biochemistry has acquired an individuality of its own. A biochemist to-day, one might predict with some confidence (though not entirely without regret), would be most likely to hold forth, among biologists or chemists alike, about concepts rather different (at first sight, at least) from the traditional ones of either biology or chemistry.

How, then, does biochemistry live up to the fusion of biology with chemistry implied by its name?

One kind of job that biochemists undertake is, of course, to isolate compounds from living things and determine their structures. In this, they share the preoccupation of other kinds of biologists with spatial form. Biochemistry includes a sort of submicroscopic anatomy that elucidates structure on the minute scale of molecules. The classical anatomists cut up bodies to describe the parts of which they are made in so far as they are visible to the naked eye. Microscopy revealed a whole new world of structure and organization smaller than this, and when the technique had developed sufficiently—when the sting had been taken out of chromatic aberration in the nineteenth century—cells became the focus of interest. With the advance of chemistry, it gradually became possible to tackle biological architecture even on the molecular scale. The grand strategy remains the same—a better understanding of living things in terms of their constituent parts. The tactics, however—the choice of techniques to apply—depend on the order of size of the parts being examined. For gross anatomy, the scalpel is appropriate, for cellular structure the microscope; for parts as small as molecules, the relevant techniques are those we call chemical—hence "biochemistry". Seen in this light, biochemistry is the logical extrapolation of dissection. The idea is epitomized in the expression "molecular biology", which is now fashionable.

Merely to determine structure, however, is far from the summit of the ambitions of biochemists. They are interested not only in what the constituents of living things are like, but also in what they do—in the way that chemical processes underlie the more obvious vital manifestations. The continuous change which is one of the most striking characteristics of life rests on unceasing chemical activity inside living organisms. Biochemistry thus continues another classical tradition of biology in linking form with function. Like anatomy divorced from physiology, static divorced from dynamic biochemistry is too dry for most intellectual palates; it is also poor in practical application, and thus fails to measure up to the Baconian ideal for science—that it should increase man's power over nature. Life, after all, is a matter of keeping events going, not only of maintaining structures; and biochemists seek to elucidate events as well as structures by isolation.

By and large, then, while the techniques of biochemistry are chemical, its problems are the basic ones of biology. Chemistry is its means, biology its end. It is the extreme extension of that approach to the phenomena of life which seeks to explain them in terms of the sub-units of which living organisms are composed. Of this kind of biological analysis, it represents the ultimate stage—ultimate because pushing the analysis a stage further, from the molecular level down to the atomic, leaves no characteristically biological kind of organization, the atoms being the same as in the inorganic realm. In so far as parts are more fundamental than wholes, biochemistry can thus claim to be the most fundamental of the biological sciences.

Enthusiasts for other kinds of approach to biological problems need take little umbrage at this claim, which does not amount to an arrogant pretension to be the most judicious, the most useful or the most illuminating. By the same token that makes it the deepest form of biological analysis, biochemistry also concentrates on the furthest remove from immediate biological reality; in so far as it concerns itself with molecules, it remains remote from intact organisms. Data on the molecular level have to be related to observations made on more highly organized, less disrupted systems. The whole question of correlating observations made at different levels of organization, a major issue for experimental biology in general, is particularly crucial for biochemistry, and will rear its intriguing head again later (pp. 95–98).

For the moment, however, it is time to leave the airy realm of philosophical generalities and come down to the hard bedrock of particular facts. What are the molecular parts of living organisms which are the special concern of biochemistry? Three main groups of them—proteins, carbohydrates and fats—have, in their roles as major constituents of a balanced human diet, become matters of general knowledge. (No woman's magazine nowadays can go more than a few issues without reminding its readers of them in connection with some new slimming diet.) A fourth group, the nucleic acids, though it has not achieved the same nutritional notoriety, is nevertheless universally distributed in all forms of life.

Is it possible to pick out any of these groups as being more indispensable than the others? Scientists disillusioned by the vagueness of the old concept of "protoplasm" as the "physical basis of life" sometimes try to do so. Making a choice is not easy. It soon becomes like trying to decide which is more important for a motor-car, the engine or the wheels; the point is that no contraption lacking either would be recognized as a motor-car.

On behalf of proteins, it can nevertheless be said that they in particular play a key role by virtue of the fact that the enzymes, which catalyze virtually every reaction in living organisms, are all proteins. This makes them, at least, a good topic with which to begin a discussion of biochemistry—a topic which also illustrates some of the problems of grappling with the sort of large molecules with which biochemists have perforce to concern themselves. For these reasons, it is to proteins that the next chapter is devoted.

II

Big Molecules—The Proteins

Large and labile—handle with care! Purifying proteins

PROTEINS are built of units which, having both amino and carboxyl groups, are called amino-acids and have the general formula $NH_2 \cdot CHR \cdot COOH$, where R stands for a group which is different in each amino-acid. The carboxyl group of one amino-acid is linked in proteins to the amino group of the next in an amide or peptide bond. This linking can be formally represented as involving the elimination of the elements of water:—

$$-COOH + NH_2- \rightarrow -CO-NH- + H_2O$$

Clearly, such linking can in theory be continued indefinitely to give peptide chains of any length. Although only about 20 different kinds of amino-acid occur generally in proteins (Table I), many

TABLE I. *Amino-acids in proteins*

Name	Abbreviation	Characteristic feature
Glycine	Gly	
Alanine	Ala	
Valine	Val	R is simple aliphatic group
Leucine	Leu	
Isoleucine	Ileu	
Serine	Ser	R contains hydroxyl group, —OH
Threonine	Thr	
Cystine	CySSCy	
Cysteine	CySH	R contains sulphur
Methionine	Met	
Aspartic acid	Asp	R contains carboxyl group, —COOH
Glutamic acid	Glu	
Asparagine	Asp·NH_2	R contains amide group, —$CONH_2$
Glutamine	Glu·NH_2	
Lysine	Lys	
Arginine	Arg	R contains various basic groups
Histidine	His	
Phenylalanine	Phe	
Tyrosine	Tyr	R contains aromatic ring
Tryptophan	Try	
Proline	Pro	>NH instead of —NH_2

BIG MOLECULES—THE PROTEINS

residues of any one kind may be present. It is found, in fact, that proteins have very large molecular weights, ranging from several thousands to several millions, which means that the numbers of amino-acid residues in the molecules of different proteins lie between about fifty and some tens of thousands.

Notes to Table I.

Formulae. The general formula for the amino-acids is $NH_2 \cdot CHR \cdot COOH$, R being different in the various amino-acids. (Proline is an exception, having an imino group, $>NH$, instead of the amino group.) Glycine, $NH_2 \cdot CH_2 \cdot COOH$ ($R = H$) is the simplest amino-acid; next comes alanine, $NH_2 \cdot CH(CH_3) \cdot COOH$ ($R = CH_3$). In the third column of the table are shown characteristic features of other R groups.

The R groups of aspartic and glutamic acids are acidic, containing a carboxyl group (additional to the α-carboxyl group common to all amino-acids). If the second carboxyl group is converted into an amide, the resulting compounds are asparagine and glutamine, respectively. Both the acid and the amide forms occur in proteins.

$$\begin{array}{cccc}
COOH & CONH_2 & COOH & CONH_2 \\
| & | & | & | \\
CH_2 & CH_2 & (CH_2)_2 & (CH_2)_2 \\
| & | & | & | \\
NH_2CHCOOH & NH_2CHCOOH & NH_2CHCOOH & NH_2CHCOOH \\
\text{Asp} & \text{Asp} \cdot NH_2 & \text{Glu} & \text{Glu} \cdot NH_2
\end{array}$$

Cysteine contains a sulphydryl group, $-SH$. On oxidation, sulphydryl compounds unite in pairs to give disulphides:—

$$2X-SH \xrightarrow{-2H} X-S-S-X$$

Cystine is the disulphide form of cysteine; it is, in effect, a double amino-acid (cf. p. 35)

$$\begin{array}{ccc}
SH & S\text{------} & \text{------}S \\
| & | & | \\
CH_2 & CH_2 & CH_2 \\
| & | & | \\
NH_2CHCOOH & NH_2CHCOOH & NH_2CHCOOH \\
\text{CySH} & & \text{CySSCy}
\end{array}$$

Some further amino-acid formulae are given under Table II.

Number of different amino-acids in proteins. The number depends on whether closely related pairs are counted separately or together. It is best, perhaps, to take it as twenty—not only because this is a nice round number, but because this seems to be the number of different units whose arrangement in sequence is determined by the mechanism controlling protein synthesis (p. 175). It is arrived at by counting cystine and cysteine together, but the acidic amino-acids (aspartic and glutamic acids) separately from their amides. (During hydrolysis, of course, the amide groups are split off, so that the asparagine and glutamine of proteins turn up as the corresponding acids in hydrolysates.)

The twenty amino-acids listed are the only *general* protein constituents; a few other amino-acids occur in some special proteins.

Largely because they have such enormous molecules, proteins are different in many ways from the sort of compound with which an organic chemist feels at home. Classical organic chemistry relies largely on recrystallization for purifying solids, and on melting points for testing purity. Proteins when heated do not melt but only char. Many of them have been crystallized, but more have not; even when crystallization is possible, it is no sure way of purifying proteins, since they readily form mixed crystals with each other.

The situation is not made easier by the considerable lability of proteins. Handling them calls for a far gentler approach than is usual for more typical organic molecules. Boiling in water, or exposure to quite dilute acid or alkali—one-tenth normal, say—are treatments too rough for proteins (with rare exceptions) to withstand without suffering damage by a process known as denaturation.

How, then, can proteins be purified from the complex mixtures in which they occur biologically?

One of the best-tried methods is precipitation with ammonium sulphate. This salt is exceedingly soluble, 100 g of water being capable of dissolving 76 g of it at room temperature. Proteins are insoluble in the saturated solution. As the ammonium sulphate concentration in a protein solution is raised, one protein after another is precipitated or "salted out". Different proteins come out of solution at different degrees of saturation with ammonium sulphate, and can thus be separated from each other. Sometimes the precipitates can be collected by filtering, but often they filter badly and it is better to spin them off in a centrifuge; such instruments, with capacities ranging from a few millilitres to several litres, are basic equipment for biochemical laboratories.

The products can be freed of ammonium sulphate by dialysis. For this simple operation, they are put in a length of cellophan dialysis tubing which is closed by tying off at both ends and put into a large volume of water for some hours, preferably with stirring. The tubing acts as a kind of molecular sieve; its pores are large enough to allow the salt to diffuse out, but small enough to retain the big protein molecules. By replacing the outside liquid with fresh water several times, all dialysable material can be removed.

As an alternative to ammonium sulphate, alcohol has often been used as precipitant. In this case, the solution has to be kept cold (near $0°C$) to avoid damaging the protein by denaturation. (Further methods for purifying proteins are discussed later—pp. 41–47.)

BIG MOLECULES—THE PROTEINS

Fibrinogen from blood, and ovalbumin from egg white

These general procedures can be illustrated by some particular examples, taking first fibrinogen, the protein in plasma which, by turning into insoluble fibrin, causes blood to clot (p. 51). This protein is rather easy to prepare from ox blood. To prevent clotting, the blood must be mixed, immediately on collection at the slaughterhouse, with an anticoagulant such as sodium citrate. In the laboratory, the blood corpuscles are first separated from the plasma by centrifuging. The corpuscles can be thrown away (or taken home as a fertilizer for the rose-bed, for which purpose they are excellent); it is the plasma, the pale, orange-coloured supernatant liquid, which contains the fibrinogen. Many other proteins are also present in it (Fig. 9), but fibrinogen is among the first to precipitate on addition of either ammonium sulphate or alcohol. Of the protein in the white precipitate formed by bringing the alcohol concentration to 8% by volume, about 70% is fibrinogen; while 20% saturation with ammonium sulphate gives fibrinogen of a purity as high as 90%. In each case, further purification can be achieved by repeated precipitations.

Another illustrative example is ovalbumin, the major constituent (apart from water) of egg white. Unlike fibrinogen, this protein is quite easy to crystallize. Ammonium sulphate is added to egg white to 50% saturation and the precipitate, which contains some of the other proteins, is removed. The solution is now made slightly acid (pH 4·8) with dilute sulphuric acid; this must be done carefully, with gentle but steady stirring, since excess of acid denatures the protein. More ammonium sulphate is now added until a permanent turbidity shows that the ovalbumin is on the verge of precipitation; on allowing to stand a day or two, it crystallizes as microscopic needles. When stirred, the suspension exhibits a beautiful silvery sheen as the tiny needles orient themselves in the direction of flow. (The relatively enormous crystals which may separate where the edges of the solution dry out are often taken for protein by optimistic novices, but they are only ammonium sulphate. Even experienced workers have been known to fall into similar or even more humiliating traps, such as mistaking air-bubbles for protein crystals.)

Protein "heat-stroke"—denaturation

One obvious way in which this crystallization differs from the sort usual in organic chemistry is that it does not involve heating. There is a very good reason for this. Proteins are sensitive, and if the temperature rises too high, they are damaged—they suffer "heat-stroke", so to speak. The process by which they are damaged is known as denaturation; it has already reared its ugly head several times, and more must now be said about it.

The denaturation of ovalbumin by heat is in fact a process familiar to everyone, because it occurs when an egg is boiled. Its most obvious feature is that the protein is turned into a less soluble form. The whole solution, being in this case so concentrated (about 10%), sets to a gel; more usually, on boiling a dilute protein solution, the denatured product separates as a flocculent precipitate. Organic solvents denature many proteins even at room temperature. Thus, if care is not taken during alcohol precipitation to keep the temperature down, denaturation often occurs and attempts to redissolve the protein in water or dilute salt solution are in vain. Acid and alkali of quite moderate strengths also cause denaturation, although in these cases no precipitate usually appears, since even denatured proteins are frequently soluble in acid or alkaline media (p. 40); on adjusting the pH back to near neutrality, however, the material often separates, now insoluble under conditions in which previously it was soluble.

The circumstances which bring about denaturation are so mild that they could hardly break ordinary chemical bonds of the covalent type. What seems to happen to proteins during denaturation is that, while the chemical structure of the amino-acid residues and the peptide bonds between them remains unaffected, the folding of the long peptide chains is altered. In native proteins, the chains do not flop about in random fashion but are folded in specific configurations. These folds are held in place largely by such forces as hydrogen bonds (Fig. 23), which are considerably weaker than ordinary covalent bonds and could be disrupted by such treatments as heating in water. When the protein suffers "heat-stroke", the native configuration of its chains, precariously held by relatively feeble forces, just collapses.

Another way of bringing about denaturation is by stirring or whisking a protein solution into a foam. That is why, although efficient mixing is very necessary during the addition of sulphuric acid in the crystallization of ovalbumin, it must not be so violent as to work up a great froth. This type of denaturation is called surface-denaturation, because it seems to be caused by surface tension pulling molecules of protein at the air-liquid interface out of the native configuration. When fresh interface is constantly reformed, appreciable amounts of protein eventually become denatured. The surface-denaturation of ovalbumin is almost as well known in everyday life as its heat-denaturation, because it occurs when egg white is beaten until stiff. In making meringues, the sweetened egg white is first surface-denatured by whisking and the process is then finished by heat in the oven (although good cookery books say that the heat should be very gentle, to dry out rather than bake).

Decrease in solubility is often the most obvious change in properties on denaturation, but it is not the only important one. Some groups in native proteins often fail to show their normal chemical reactivity because the folding of the chains is such that they are tucked away beyond reach of reagents in the middle of the molecule. With the unfolding that accompanies denaturation, they become accessible and react as expected. For instance, the red colour with alkaline sodium nitroprusside, which depends on the —SH groups of cysteine residues (Table I), is often weaker with a native than with the corresponding denatured protein. Similarly, denaturation increases the susceptibility to hydrolysis by proteolytic enzymes, such as those of the alimentary tract whose job it is to digest food proteins by splitting peptide bonds. The unfolded, denatured molecule is more exposed to attack than the folded, native one. Thus, if a suitable enzyme is allowed to act on solutions of native and denatured ovalbumin, the latter is digested faster.

Most significant, perhaps, among the changes accompanying denaturation is the loss of biological activity. Among the types of activity which different proteins exhibit are those of enzymes, of antibodies and of hormones. (There are also non-protein hormones, but all enzymes and all antibodies are proteins.) Denaturation destroys the power of enzymes to catalyse specific reactions, of antibodies to bind their specific antigens (p. 132) and of protein hormones to produce characteristic effects on their target organs. Thus, for a protein to perform its role in the body, it must not only have the right nature and number of amino-acids arranged in the right sequence; the chains must also be folded in the right specific configuration.

There are, therefore, a number of different kinds of important information to be got about the structure of a protein. There is the identification of the amino-acid units present, and the estimation of their amounts; there is the determination of the sequence in which they occur in the peptide chains; and there is also the question of the spatial arrangement of the chains in the native molecule. These points are considered in turn below.

Protein structure (i) The units in the chains—amino-acid analysis

For releasing amino-acids from their combination in proteins, the usual procedure is to heat in strong acid (e.g., 6N HCl at 105°C for 24 hours). The problem then is how to work up the relatively complex mixtures that result—to isolate the individual amino-acids and find out how much of each is present.

During the early part of the present century, a great deal of hard work was done to develop procedures for quantitatively isolating and weighing the individual amino-acids—as sparingly soluble derivatives that can be crystallized, for instance, or as volatile ones that can be fractionally distilled under reduced pressure. Nowadays this approach is, in general, as extinct as the dodo, and it is sad, in a way, to think of all the devoted effort and first-class ability that was put into it. In the years before the Second World War, a complete amino-acid analysis of a protein was a matter of several months' work, and demanded a degree of manipulative skill which few chemists to-day ever bother to acquire. Chemistry has not been immune from the general decline of craftsmanship that accompanies the introduction of more efficient methods.

What has changed the whole outlook is, of course, the development of chromatographic methods. Their advent has so profoundly affected the whole way of life in biochemical (and other) laboratories that it is worth making a considerable digression to consider them here.

A digression on chromatography—separations in columns and on paper

Chromatography is essentially a method of separation. Substances are separated by virtue of differences in the rates at which they move in the chromatographic system, which consists of a mobile phase moving over a stationary one. The substances undergoing separation—typically members of a series of similar compounds—are carried along by the moving phase, but also held back to varying extents by their affinities for the stationary phase; according to the ratio of their affinities for the two phases, they move faster or slower and can thus be separated if the ratios differ. In what is now the most popular kind of chromatographic system, paper chromatography, a sheet of filter paper is the stationary phase and a solvent soaking along it the mobile phase. Different compounds are washed along the paper at different rates and therefore take up different positions on the paper. In column chromatography, a powdered solid packed into a tube takes the place of the paper and a mobile phase is made to flow through the resulting column.

Biochemists can claim much of the credit for introducing chromatographic methods. Although often they have taken over and applied to their own ends techniques which other kinds of chemists had introduced for non-biological uses, the reverse process has not been as uncommon as many suppose. For methods of separation, in particular, much initiative and impetus has come from workers interested in biological materials. It is these who, making a virtue of necessity, have been most ready to get to grips with the problems of fishing individual components out of the most complex and unpromising mixtures. ("*Tierchemie ist nur Schmierchemie*" runs an old German saying, meaning that animal chemistry is only the chemistry of slimes and messes, and few would deny this jibe a certain amount of justification.) Classical organic chemists, by contrast, have often taken pride in avoiding rather than tackling the difficulties of isolation from complex mixtures. Knowing that a minor impurity in a starting material often leads to a major impurity in a product, they have made it a matter of professional self-respect to use only substances of high purity. The cleaner a reaction mixture, the higher is the chance of crystallizing the product; in a long synthesis involving many steps, therefore, it has usually been better to go to considerable trouble to purify each intermediate, rather than to dash headlong on to the end of the series of reactions and then try to extract the desired needle from the haystack of contaminants.

It is thus more than mere coincidence that the major advances in chromatography have been made in connection with work on biological materials. The introduction of the technique in its column form was due to Tswett, a botanist investigating plant pigments, in what is now Poland but was at the time concerned, in 1906, part of Russia. That form of the technique which shows it at its most elegantly simple, paper chromatography, is due to Consden, Gordon and Martin, who were investigating keratin, the protein of wool, at Leeds in 1944; they used it for the very purpose with which we are here concerned—namely, the separation of the amino-acids in protein hydrolysates.

What Tswett did was to allow a solution of green leaf pigments in an organic solvent (petroleum ether) to run through a column of calcium carbonate powder tightly packed in a vertical glass tube. The calcium carbonate acted as stationary phase, the organic solvent as moving phase. Because the various pigments were adsorbed with different strengths by the calcium carbonate, they separated into a series of coloured zones, the most strongly adsorbed remaining nearest the top. "This separation becomes virtually complete" wrote Tswett in his 1906 paper "if, after the pigment solution, pure solvent is passed through the column. Like rays of light in the spectrum, the different components of a pigment mixture are separated according to their properties in the calcium carbonate column and can thereby be qualitatively and quantitatively determined. Such a preparation I call a chromatogram, and the corresponding method, the chromatographic method". (Chromatography derived its name from this early use with pigments; it is not, of course, restricted in its application to coloured substances.)

Procedures similar to Tswett's were used a certain amount during the following decades, but suitable adsorbents were available for only a few types of compound. It is easy enough to find adsorbents to adsorb most substances, but only rarely is the adsorption freely reversible. Once on the adsorbent, the substances often stick tight and refuse to be washed off cleanly into a liquid. Mainly because of this snag, it was not until the nineteen-forties that chromatography became the technique of almost universal applicability that it is to-day.

In a paper published in the *Biochemical Journal* in 1941, Martin and Synge described how they had substituted for a solid adsorbent a stationary phase of liquid held immobile on an inert solid. Instead of adsorbing on a solid, the substances to be separated now partitioned between two liquids, one mobile, the other stationary. Unlike adsorption, partition is in general freely reversible, and the ingenious idea of partition chromatography—as distinct from adsorption chromatography—found so many uses that Martin and Synge were jointly awarded the Nobel prize for chemistry in 1952. What they themselves did was to mix silica gel powder with about half its weight of water and use this moist gel to pack columns similar to Tswett's, through which they ran organic solvents immiscible with water (e.g., chloroform containing 1% n-butanol). On such columns, they separated mixtures of amino-acids in the form of their acetyl derivatives, which ran at different rates according to their relative affinities for water and the organic solvent (according to their partition coefficients, in other words). As the column was washed with more and more solvent, the amino-acid derivatives emerged one by one from the bottom of the column, those with the greater affinities for organic solvent relative to water emerging earlier (Fig. 1).

Paper chromatography developed as an offshoot from partition chromatography in columns. The idea occurred to Consden, Gordon and Martin to dispense altogether with a column packed in a glass tube and to use instead a plain sheet of filter paper, the cellulose fibres of which in a moist atmosphere always bind a certain

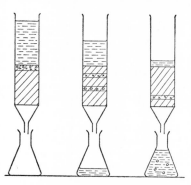

Fig. 1. *Column chromatography*
The stationary phase (solid adsorbent for adsorption chromatography, or inert solid holding liquid for partition chromatography) is made into a slurry with the solvent to be used as mobile phase and packed into the column, making sure that packing is even and that no air bubbles are included, so that flow through the column is uniform. The sample, represented here as a mixture of noughts and crosses, is applied to the top of the column in a little solvent. More solvent is then made to run through the column to develop it; the noughts and the crosses are washed down the column as bands moving at different rates.

amount of water to serve as stationary phase. The mobile phase used to develop the chromatogram was merely allowed to soak along the paper.

This form of chromatography is really extraordinarily simple. It may seem impolitic nowadays to trumpet abroad too loudly the successes achieved by scientific ingenuity with the cheapest of means; senior scientists already have to spend much time begging Government or industry for money to buy equipment (indeed, it sometimes seems that degree of seniority can be measured by the proportion of time spent begging instead of in the more direct pursuit of science). Nevertheless, it must be admitted as an arguable proposition that so simple a technique as paper chromatography would never have occurred to workers in a lavishly stocked laboratory. The invention came at the height of the Second World War, when shortages of all kinds were severe, and there may be a germ of truth in the idea that nobody would have started playing with bits of filter paper if more sophisticated apparatus and materials had been available. Perhaps the invention of paper chromatography was a genuine case of genius flourishing in adversity.

The essentials of a paper chromatographic system really are laughably easy to set up. All one needs for a demonstration at home is to put a spot of red ink near one edge of a piece of filter or blotting paper, and then dip the edge into water; as the liquid soaks past the spot, some makes of red ink are soon resolved into differently coloured components. (The fact that separation can be obtained with an aqueous mobile phase shows, incidentally, that the simple idea of partition between two liquid phases, one of them stationary water, does not explain all the processes at work in paper chromatography.) The apparatus in common laboratory use is not much more complicated, except that the paper is hung inside a closed tank, and the solvent, instead of ascending, is usually made to flow downwards from a trough into which the top end of the paper is dipped. Even suppliers of scientific equipment, who sometimes try to add a little gadgetry when the essentials of an apparatus seem too straightforward to sell well, have failed to raise prices beyond modest levels. Commercially, the main beneficiaries have probably been the filter paper manufacturers—who, recovering from their surprise at such unexpected good luck, have marketed sheets specially selected for chromatographic use.

Paper chromatography spread like wildfire through the laboratories of the world. By the end of the nineteen-forties, almost every conceivable type of substance that can be made to dissolve was being "run on paper", mostly with success, in systems which hardly differed in essentials from the original except in employing new solvents and solvent mixtures. The sort of purpose for which paper chromatography has proved itself unrivalled is to detect components of mixtures and identify them by comparison with known substances placed as "markers" on the paper by the side of the unknowns (Fig. 2); identical rates in several different solvent systems constitute evidence as good as any that two materials are the same. Chromatography can also be used as a test of purity, for failure to detect more than one component is evidence that only one kind of molecule is present.

Besides economy of expense, the main advantages of paper chromatography are economy of effort and economy of material. More than a dozen unknowns and markers can be put on a single sheet

BIG MOLECULES—THE PROTEINS

of paper. Often it is convenient to allow the solvent to run overnight, and by 11 o'clock next morning the paper can be dried and sprayed. (The research worker can thus have fresh food for thought over his morning coffee, and the old excuse for late rising—that no discovery was ever made before lunch—has unfortunately become feebler than ever.) The amount of material necessary depends only

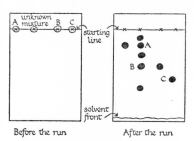

Before the run   After the run

Fig. 2. *Paper chromatography—use of markers*

Before the run, a solution of the mixture of unknowns (e.g., amino-acids) is applied as a small spot near one edge of the sheet of filter paper. Spots of the known compounds A, B and C are also placed on the starting line to act as markers. The paper is developed by dipping the top edge into a trough of the solvent mixture chosen as mobile phase, and allowing it to soak down as far as the position marked "solvent front". (For amino-acids, a well-tried mixture is the upper phase formed after mixing n-butanol, acetic acid and water in the proportions 4:1:5 by volume.) Each component of the mixture moves down the paper a characteristic distance relative to the distance run by the solvent, and thus comes to form a separate spot. To locate these spots, the paper after drying is sprayed with a solution of a reagent which forms a colour with them. (For amino-acids, the commonest such reagent is ninhydrin, which on heating gives a purple colour.)

Two of the components of the mixture can be presumptively identified with the knowns A and B. To check the identification, further chromatograms can be run with different solvent systems. C is not present in the unknown mixture—at least, not in a concentration comparable to those of the five major components. To characterize the three as yet unidentified spots, more markers must be used in subsequent runs.

on the sensitivity of the means used to make the spots visible at the end of the run. With the ninhydrin spray used for amino-acids, amounts of a few micrograms (1 microgram is one millionth of a gram) can easily be detected. When radioactive substances are used, smaller quantities still can be located on the paper. Smallness of the amounts of material handled is, indeed, a limitation of the versatility of the method, which is difficult to scale up beyond the milligram level for preparative purposes.

In studying the composition of proteins and their derivatives, paper chromatography has been invaluable for the qualitative purpose of determining which amino-acids are present. It can also be made semi-quantitative, but for accurate analyses column chromatography is better. The most successful system in recent use employs columns packed with ion exchange resins (p. 41).

Once the amino-acids are separated, it is quite easy to estimate their amounts; usually, this is done by measuring the amount of colour formed on heating with ninhydrin. The colour intensity of a solution is determined in a spectrophotometer, a very popular kind of apparatus in biochemical laboratories, which measures the amount of light of a selected wavelength that is absorbed on passing through a known length of solution. To estimate the amount of amino-acid in a solution, the colour intensity it gives is compared with those given by known amounts of amino-acid heated under the same conditions with ninhydrin. Quantities of as little as a few micrograms can be estimated quite accurately by this method.

Protein structure (ii) Order in the chains—amino-acid sequence. Insulin

Most proteins contain the full complement of the usual 20 different amino-acid sub-units. Their proportions vary, of course, but often not very drastically between one protein and another. It is in the order in which the amino-acid residues are arranged that the significant differences between proteins lie.

There are, of course, a vast number of possible arrangements. Everybody is aware that there can be many permutations of a few different kinds of units, but most people have not grasped just how astronomically numerous the possibilities are. (On precisely this fact are founded the fortunes made by football pool promoters.) The arithmetic is, however, not difficult to do. Taking the units in pairs (i.e., considering dipeptides, each containing two residues), there are 20 possibilities for the first position and 20 for the second, making 20 × 20 or 400 possibilities in all. For tripeptides, there are again 20 possibilities for filling the third position, so that the total number of possibilities is 20^3 or 8,000. In the general case, for a peptide containing n residues altogether, the number of arrangements is 20^n. When n is only 10, this already works out at more than ten million million. But proteins contain hundreds or thousands of residues per molecule. Even ignoring all cases with more than a thousand residues (i.e., with molecular weights above about 100,000), there is nowhere near enough matter in the earth to make a single specimen molecule of every possibility. It is thus more than a mere empty figure of speech to say that the number of possible proteins is infinite for all practical purposes. Such is the versatility of the ground-plan on which Nature has chosen to build living organisms; for the diversity of protein molecules is the molecular basis of the great variety among the living forms we see around us.

These thoughts are impressive from the standpoint of the general philosophy of biology, but formidable and discouraging for the experimenter hoping actually to determine sequences. Apart from some over-optimistic guesses that particular types of residue might recur at regular intervals along the chain, the problem seemed quite intractable up to the time of the Second World War. Without chromatography and related techniques (p. 45), it would still be so. The man who showed that, given such techniques, it could be tackled with success was Sanger, of Cambridge. Between the mid-nineteen-forties and the mid-nineteen-fifties, working at first alone and later with a small handful of collaborators, he elucidated the amino-acid sequence of insulin (the protein hormone of the pancreas which is deficient in diabetes and is used to treat the disease). For this, the first-ever determination of the complete structural formula of a protein, he was awarded the Nobel prize for chemistry in 1958.

Every peptide chain must, of course, have two ends (unless it forms a closed loop), and to identify the amino-acid residues occupying the terminal positions is a first step towards ascertaining the complete sequence. Chain ends are of two types; at one, an amino group remains free, at the other a carboxyl group. In the case of non-terminal residues, both amino and carboxyl groups are blocked by peptide bond formation with neighbouring residues.

$$NH_2 \cdot CHR^1 \cdot CO \underbrace{\hspace{2em}}_{\text{amino-terminal residue}} NH \cdot CHR^2 \cdot CO \underbrace{- - - - - NH \cdot CHR^{x-1} \cdot CO}_{\text{non-terminal residues}} \underbrace{NH \cdot CHR^x \cdot COOH}_{\text{carboxyl-terminal residue}}$$

Sanger's first major contribution was to develop a good method for determining amino-terminal residues (i.e., those which bear free amino groups). In his method, the protein or peptide is first treated with fluorodinitrobenzene, $C_6H_3(NO_2)_2F$, which under mild conditions reacts with free amino groups to give yellow dinitrophenyl derivatives. Fluorodinitrobenzene can be thought of as dinitrophenyl fluoride, and writing DNP for the dinitrophenyl group, $C_6H_3(NO_2)_2—$, the process can be represented as follows:—

$$DNP—F + NH_2—X \xrightarrow[\text{room temperature}]{\text{aqueous NaHCO}_3} DNP—NHX + HF$$

Next, the treated protein or peptide is heated in strong acid to hydrolyse all the peptide bonds. Most of the amino-acids are liberated in the free state, but those which were originally amino-terminal are left as DNP derivatives.

$$DNP \cdot NH \cdot CHR^1 \cdot CO\text{—}NH \cdot CHR^2 CO\text{-------}$$
$$\downarrow \text{6N HCl, 105°C}$$
$$DNP \cdot NH \cdot CHR^1 \cdot COOH + NH_2 \cdot CHR^2 \cdot COOH \text{ etc.}$$

The DNP group thus "tags" or "labels" the amino-terminal position, and is doubly convenient for this purpose because its yellow colour makes it easy to follow the chromatography of DNP derivatives and to estimate their amounts. Identification is achieved, of course, by comparing DNP derivatives from proteins with known, synthetic ones by means of chromatography (that is, by comparing their rates of movement in various chromatographic systems). It is quite easy to put the results on a quantitative basis by measuring the intensity of the yellow colour of a solution in a spectrophotometer (p. 30).

From insulin, Sanger got DNP derivatives of glycine and of phenylalanine, indicating that the molecule contains peptide chains of two different kinds (cf. p. 35), ending in these two amino-acids, respectively. Further information was obtained by submitting DNP-insulin to milder acid treatment so as to split only some of the peptide bonds and leave a mixture of small peptides, each still containing a few amino-acid residues. Such partial hydrolysates contained more different yellow DNP compounds, which were also separated by chromatography. One of these gave, on complete hydrolysis, free isoleucine and DNP-glycine; its structure must, therefore, be DNP-glycyl-isoleucine, and isoleucine must have been the penultimate residue in the intact chain of insulin. (In the formulae above, $NH_2 \cdot CHR^1 \cdot COOH$ would in this case stand for glycine, and $NH_2 \cdot CHR^2 \cdot COOH$ for isoleucine. Another yellow compound gave valine in addition to isoleucine and DNP-glycine on complete hydrolysis, thus establishing the sequence Gly.Ileu.Val. .... By such means, amino-acids were allocated to the four or five positions nearest the amino-terminal ends of the chains.

This work on terminal sequences was impressive, but most experts could not help feeling dubious about the prospects of carrying such an approach much further to cover the whole length of the chains. To Sanger, however, having got so far, the goal no longer seemed unattainable. Undaunted, no longer restricting his interest to chain ends, he broke down insulin in a number of ways, mostly by mild acid hydrolysis or by proteolytic enzymes. Literally scores of peptides from the resulting partial hydrolysates were separated on paper; such a project, of course, would never have got beyond the stage of a day-dream in the days before chromatography. The amounts of each peptide isolated were often not even on the milligram scale, but sufficed to identify the amino-acids present by paper chromatography, and to determine which of them was amino-terminal by the DNP technique. Gradually, as the results accumulated, it became possible to fit together larger and larger pieces of what amounted to a molecular jig-saw puzzle.

To see how the fitting together was achieved, suppose that among the peptides isolated was one containing the amino-acids A, B and C, A being amino-terminal; that another contained A, B, C, D and E, A being again amino-terminal; and that a third contained only D and E, D being amino-terminal. These would suggest the sequence to be either A–B–C–D–E or A–C–B–D–E. The former structure would be indicated and confirmed if among the breakdown products were found other peptides containing, for example, A and B (A amino-terminal), or B and C (B amino-terminal), or B, C and D (B amino-terminal) or C, D and E (C amino-terminal). In practice, of course, the situation was much complicated by the fact that more than one residue of most of the amino-acids was present. Eventually, however, a unique sequence was arrived at, and the complete structure of insulin was published in a paper in the *Biochemical Journal* in 1955 (Fig. 3).

To avoid giving a misleading impression of the situation in protein chemistry, it must be emphasized that the success with insulin, great though it was, remains relatively isolated. It is still too early to say that the problem of determining amino-acid sequences is all over bar a lot of routine work with well-established methods. The molecular weight of insulin is, at 6,000, much lower than that of most proteins. Although the amino-acid sequences of a few somewhat larger molecules are now known, those of the vast majority of proteins still await elucidation.

Protein structure (iii) Folding of the chains

The way in which the peptide chains of native protein molecules are folded is another structural feature about which the first snatches of information have only recently begun to appear. The problem falls outside the terms of reference of ordinary chemical methods. It has been approached by analysing the diffraction patterns obtained by shining X-rays at protein fibres or crystals; some rather complicated mathematics figures in the theoretical background to such work,

Fig. 3. *Structure of beef insulin*

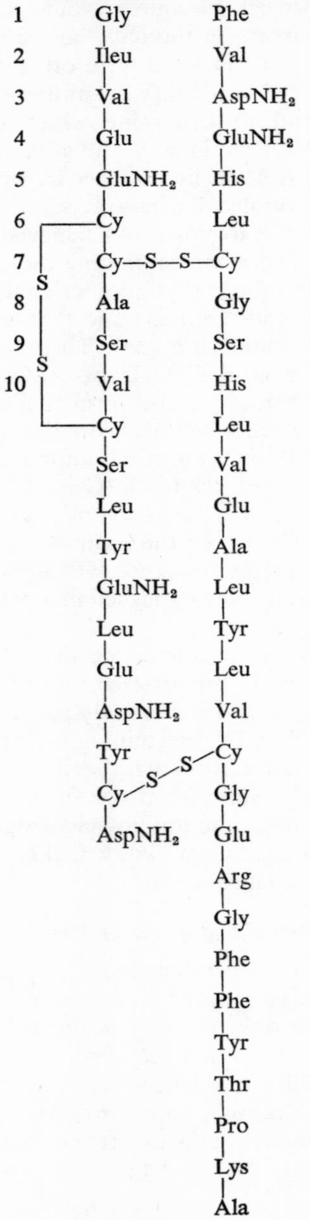

1		Gly	Phe
2		Ileu	Val
3		Val	AspNH₂
4		Glu	GluNH₂
5		GluNH₂	His
6		Cy	Leu
7		Cy—S—S—Cy	
8	S	Ala	Gly
9		Ser	Ser
10	S	Val	His
		Cy	Leu
		Ser	Val
		Leu	Glu
		Tyr	Ala
		GluNH₂	Leu
		Leu	Tyr
		Glu	Leu
		AspNH₂	Val
		Tyr	Cy
		Cy—S—S—Gly	
		AspNH₂	Glu
			Arg
			Gly
			Phe
			Phe
			Tyr
			Thr
			Pro
			Lys
			Ala

and the computing involved is such as to call for advanced electronic computers. Perutz and Kendrew, of Cambridge, were awarded the Nobel prize for chemistry in 1962 for their work in this field.

Protein structure (iv) Cross-linking the chains—disulphide bridges. Keratin

The molecules of most proteins have not one but several peptide chains. Cross-linking between the chains is brought about by the disulphide bonds, —S—S—, of cystine residues. Cystine is in effect a double amino-acid (Table I), and if its two halves form part of two different chains, it forms a bridge holding those chains together. Insulin, for example, has two different chains (as indicated by the presence of two different amino-terminal groups, p. 32); they are held together by two disulphide bridges, formed between positions 7 and 20 of the glycyl chain on the one hand, and positions 7 and 19 of the phenylalanyl chain on the other (Fig. 3).

In keratin, the protein of hair and wool, the cystine content is particularly high (14% in human hair), and the large number of disulphide bridges cross-linking the whole structure contributes to its strength and insolubility. The theory of the "home perm" (as distinct from permanent waving by the application of heat) is a neat piece of applied biochemistry based on attacking the disulphide bonds. Of the two solutions used, the first contains a reducing agent

---

Notes to Fig. 3.

Besides beef insulin, Sanger and his co-workers examined insulins from various other species, with the interesting result that they all turned out to be identical except at positions 8, 9 and 10 of the glycyl chain. Here, Thr sometimes takes the place of Ala, Gly of Ser and Ileu of Val. The sequences in various species are as follows:

	Beef	Pig and whale	Sheep	Horse
8	Ala	Thr	Ala	Thr
9	Ser	Ser	Gly	Gly
10	Val	Ileu	Val	Ileu

It has long been known that corresponding proteins from different species are slightly different—that proteins are species-specific, in other words. These results on various insulins gave the first precise information on this point in terms of structural formulae. The identity between pig and whale insulins is probably a matter of chance, and there seems to be no justification for reading any deep phylogenetic or evolutionary significance into it.

and the second an oxidizing agent. The reduction splits some of the disulphide bridges to give pairs of —SH groups, thereby loosening the whole structure to some extent.

$$
\begin{array}{c}
----\text{NH·CH·CO}---- \\
| \\
\text{CH}_2 \\
| \\
\text{S} \\
| \\
\text{S} \\
| \\
\text{CH}_2 \\
| \\
----\text{NH·CH·CO}----
\end{array}
\quad
\underset{\underset{-2\text{H}}{\text{oxidation}}}{\overset{\overset{\text{reduction}}{+2\text{H}}}{\rightleftarrows}}
\quad
\begin{array}{c}
----\text{NH·CH·CO}---- \\
| \\
\text{CH}_2\text{SH} \\
\\
\\
\\
\\
\\
\text{CH}_2\text{SH} \\
| \\
----\text{NH·CH·CO}----
\end{array}
$$

After the hair has been set in a way that happens currently to be in fashion, oxidation forms new disulphide bonds from pairs of —SH groups that find themselves near each other, helping to fix permanently the new (and, it is to be hoped, more desirable) configuration.

Protein molecules as packets of electric charges

A protein molecule carries both positive and negative electric charges. As far as the *net* charge is concerned, these partially or entirely cancel each other, but that does not alter the fact that both types exist as separate charges liberally dotted about the large particle.

There is nothing particularly extraordinary about opposite charges co-existing on the same molecule. Consider the case of an amino-acid such as glycine, which shows the same phenomenon in a simpler way, having only one acid and one basic group. Its formula is often written $NH_2 \cdot CH_2 \cdot COOH$, but a moment's thought makes it clear that no more than an infinitesimal proportion of it can ever exist in this form. Dissolved in water, it gives an almost neutral solution. The carboxyl group is rather as in acetic acid which, of course, can be $CH_3 \cdot COOH$ only in acid solution and becomes $CH_3 \cdot COO^-$ when neutralized with an alkali to give an acetate. Similarly, the amino group is rather like ammonia, which in the free state gives an alkaline solution and becomes NH_4^+ when neutralized with an acid to give an ammonium salt. In the glycine solution, therefore, the particles must nearly all be $^+NH_3 \cdot CH_2 \cdot COO^-$, since both amino and carboxyl groups must be predominantly charged when near neutrality. The only difference from acetic acid and ammonia is that neutralization, instead of being effected by added alkali and acid respectively, can be thought of as being due to another group in the same molecule, the H^+ for turning $-NH_2$ into $-\overset{+}{N}H_3$ coming from the $-COOH$ in turning into $-COO^-$.

If one equivalent of hydrochloric acid is added to the glycine solution, the higher H^+ concentration suppresses the ionization of the carboxyl group by a mass action effect, leaving mostly $^+NH_3 \cdot CH_2 \cdot COOH$ and an equal number of chloride ions; what has been formed is a solution of glycine hydrochloride. If an equivalent of sodium hydroxide is added to the glycine solution, the higher OH^- concentration strips an H^+ off the $-\overset{+}{N}H_3$ group, leaving $NH_2 \cdot CH_2 \cdot COO^-$ and an equal number of sodium ions; this is now a solution of sodium glycinate.

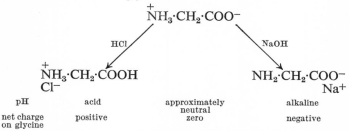

pH	acid	approximately neutral	alkaline
net charge on glycine	positive	zero	negative

Addition of acid thus titrates the acid carboxyl group, whereas addition of alkali titrates the basic amino group. This is sometimes regarded as the most amazing of paradoxes, but actually it follows quite naturally and inevitably from the state of ionization of the neutral molecule. One more point to note:—the net charge on an amino acid molecule is positive under acid and negative under alkaline conditions. The intermediate form with a net charge of zero, positive and negative exactly cancelling each other, is known as the iso-electric form, and the exact pH at which it exists is the iso-electric pH, the value of which is characteristic of the particular substance.

A slightly more sophisticated treatment of the same phenomena is in terms of pK's. The pK shows where in the pH range the change-over from undissociated form to dissociated form takes place; it is different, of course, for each particular ionizing group. At low pH, clearly, when the H^+ concentration is high, a mass action effect makes the protonated forms ($-COOH$ or $-\overset{+}{N}H_3$) predominate; at high pH, H^+ ions are released to leave the deprotonated forms ($-COO^-$ or $-NH_2$). For any group, the protonated form predominates below its pK, the deprotonated form above it (Fig. 4).

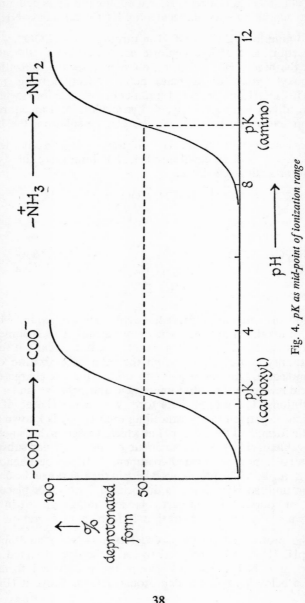

Fig. 4. *pK as mid-point of ionization range*

BIG MOLECULES—THE PROTEINS

In the particular case of glycine, there are two ionizing groups and therefore two pK's. That of the carboxyl group is 2·3, while that of the amino group is 9·6. A solution of pure glycine in water contains the iso-electric form and has a pH near 6·0. This pH is well above the pK of the carboxyl group, which is therefore almost entirely in the form of —COO$^-$; it is also well below the pK of the amino group, which is therefore almost entirely in the form of —$\overset{+}{\text{NH}}_3$.

In proteins, the ionization phenomena are essentially the same as in glycine, complicated only by the fact that each molecule has a great many ionizing groups of a number of different kinds, both acidic and basic. These have a variety of pK values spanning the whole range from quite strongly acid to quite strongly alkaline conditions (Table II).

The important general point is this. Not only is a protein always liberally sprinkled with electrostatic charges, but the pattern of charges changes appreciably with quite small changes of pH anywhere in the range between the two extremes. This is the key to

---

Notes to Fig. 4.

The p in pK denotes a negative logarithm, as in pH, and the K is the equilibrium constant for a dissociation releasing a proton, e.g.,

$$\overset{+}{\text{NH}}_3\cdot\text{CH}_2\cdot\text{COOH} \rightleftharpoons \overset{+}{\text{NH}}_3\cdot\text{CH}_2\cdot\text{COO}^- + \text{H}^+$$
$$\overset{+}{\text{NH}}_3\cdot\text{CH}_2\cdot\text{COO}^- \rightleftharpoons \text{NH}_2\cdot\text{CH}_2\cdot\text{COO}^- + \text{H}^+$$

The beauty of this way of looking at ionization phenomena is that it applies equally to acid and basic groups. It concentrates on the release of H$^+$, which is common to the two cases. One can forget about the difference—whether the H$^+$ comes from an uncharged group to leave a negatively charged one, or from a positively charged one to leave an uncharged one (i.e., whether the dissociation is AH $\rightleftharpoons$ A$^-$ + H$^+$ or $\overset{+}{\text{B}}$H $\rightleftharpoons$ B + H$^+$, where A and B stand for acidic and basic groups respectively).

The transition from protonated to deprotonated form occurs over a zone of pH, and the pK indicates, for any particular group, in which region of the pH range this zone lies. Mathematical theory shows that, when the pH of the solution is exactly the value of the pK, equal proportions of protonated and deprotonated forms are present. One pH unit above, there is ten times as much of the deprotonated as of the protonated form, and two pH units above, a hundred times. Conversely, one pH unit below the pK, there is ten times as much of the protonated as of the deprotonated form and two pH units below, a hundred times. This logarithmic relationship gives rise to the S-shaped curves in the figure; the shape is the same for all groups—only the position on the pH axis depends on the pK. The pK itself is the mid-point of the ionization range, and more than 80% of the change in ionization takes place within the zone extending one unit on either side of it.

understanding the close and sensitive dependence on pH of many properties of proteins (pp. 44, 52).

One such property is the solubility. The net charge which the molecules of a protein bear at any pH other than the iso-electric point, being of the same sign for all the molecules, causes repulsion between them which helps to keep them in solution by hindering their aggregation. The further the pH is from the iso-electric point on either side, the stronger is the mutual repulsion and the greater the solubility. At the iso-electric point, solubility is at a minimum, and if this minimum is low enough, precipitation occurs. A familiar example of iso-electric precipitation is the curdling of milk as it turns sour. Casein, the major protein of cow's milk, has an iso-electric point of 4·7 and is sparingly soluble at that pH. As the bacterial formation of lactic acid from milk sugar makes the pH drop to approach this value, the casein precipitates.

TABLE II. *Rough pK values of charged groups in proteins*

Terminal α-carboxyl	3
R groups of Asp and Glu (carboxyl)	4
R group of His (imidazole)	6
Terminal α-amino	8
R group of Lys (amino)	10
R group of Arg (guanido)	12

Since the peptide chains of proteins are usually long, there are relatively few terminal groups, and the bulk of the charge on protein molecules is, in general, contributed by acidic and basic R groups. The formulae of the acidic ones, Asp and Glu, are given under Table I; those of the three basic ones are given below. In Lys, the ionizing group is a simple amino group. His and Arg have more complex basic groups, imidazole and guanido respectively; they are shown below in the uncharged forms which predominate above the pK values

$$\underset{\text{Lys}}{\begin{array}{c} NH_2 \\ | \\ (CH_2)_4 \\ | \\ NH_2 \cdot CH \cdot COOH \end{array}} \qquad \underset{\text{His}}{\begin{array}{c} CH\!-\!N \\ \| \quad \diagdown \\ \quad \quad CH \\ \| \quad \diagup \\ C\!-\!NH \\ | \\ CH_2 \\ | \\ NH_2 \cdot CH \cdot COOH \end{array}} \qquad \underset{\text{Arg}}{\begin{array}{c} NH_2 \\ | \\ C\!=\!NH \\ | \\ NH \\ | \\ (CH_2)_3 \\ | \\ NH_2 \cdot CH \cdot COOH \end{array}}$$

BIG MOLECULES—THE PROTEINS

Handling big molecules (i) Running the electric gauntlet—ion exchange chromatography

The point has already been driven home that many of the techniques that have become stand-bys for organic chemists are not applicable to molecules as large as those of proteins. In particular, this applies to ways of purification, ways of checking purity and ways of estimating molecular size. We are now in a position to look at some of the methods of doing these things for proteins. Of the many that have been developed, it is proposed here to mention three—ion exchange, electrophoresis and ultracentrifugation.

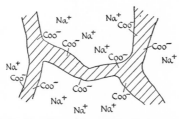

Fig. 5. *Sodium form of a carboxylic ion exchanger*

Ion exchange is a way of catching hold of molecules by virtue of the charges they carry. It thus depends on charge—as does electrophoresis, which is the movement of charged particles in an electric field. Both these techniques, therefore, can be used for charged particles, whether large or not (in practice, they are often easier to apply to small molecules). By contrast, ultracentrifugation depends on size, and is applicable to large molecules, whether charged or not.

An ion exchanger is a solid, insoluble substance bristling with ionizing groups. For instance, the groups may be carboxyl groups, which can be neutralized with NaOH to carboxylate ions. The —COO^- groups hold an equal number of Na^+ ions in the internal spaces of the ion exchanger, which is now said to be in the "sodium form" (Fig. 5). If it is washed with a solution of a potassium salt, some or all of the Na^+ ions are replaced by K^+ by a pure competition effect (no more mysterious, actually, than a double decomposition reaction between two salts in solution). The exchanger, in fact, has exchanged Na^+ for K^+; hence it is a cation exchanger. (The converse case of an anion exchanger full of positive groups is equally common and useful.)

Typically, an ion exchanger is used as the solid, immobile phase with which to pack a chromatographic column. (Its use in this way is ion exchange chromatography, as distinct from adsorption chromatography and partition chromatography (p. 26 and Fig. 1); although the processes going on in the column depend on different principles, the operations can be carried out in a similar way.) When a protein solution is run through a column of a cation exchanger in the Na^+ form, protein molecules are liable to get "caught" by the attraction of the negative groups of the exchanger for the

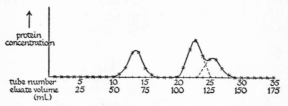

Fig. 6. *Column chromatography of a mixture of proteins*

This diagram shows a form in which the results of column chromatography are often given. After applying the protein mixture, the column of ion exchanger was washed with a salt solution of gradually increasing strength. The eluate emerging from the bottom of the column was collected in separate 5 ml. fractions in consecutively numbered test tubes. Each point in the diagram represents an estimate of protein concentration in a different tube. (The tubes, of course, replace the conical flask shown as receiver in Fig. 1.)

At first, all the proteins were held on the column, and none appeared in the first 10 fractions of the eluate. Some protein was present in tubes 11 to 16, and the symmetrical nature of the peak suggests that this is a single protein. The other peak is a double one made up of two single peaks as shown by the dotted lines. The experiment shows that at least three different proteins were present in the original mixture; presumptively pure samples of these have been obtained in tubes 11 to 16, 21 to 23 and 26 to 28 respectively. Separation of the second and third components has been incomplete, and tubes 24 and 25 contain mixtures of them.

positive groups (amino, imidazole and guanido—cf. Table II) of the protein; in other words, the protein, in its capacity as a cation, exchanges to some extent with Na^+. If the column is now washed with a dilute solution of a salt such as NaCl, Na^+ ions partially displacing the protein gradually push it down through the column or "elute" it. Different proteins are displaced with varying ease (depending on the number and disposition of the charges on them), and they can therefore be separated. One might say that the proteins are made to run the electric gauntlet of the formidable array of charged groups of the ion exchanger; some achieve this easily, others are seriously held back.

BIG MOLECULES—THE PROTEINS

Clearly, the more concentrated a salt solution, the more effective an eluting agent it is. Often, a salt solution of a given concentration will wash one protein out the column but hardly be effective at all in pushing another protein through. It is usual, therefore, to elute by means of a solution the salt concentration of which is gradually raised; this is called "gradient elution". Different proteins then emerge from the column in turn as the salt concentration reaches the different characteristic levels needed to shift them out of the electrostatic grip of the ion exchanger.

The results of such experiments are usually given in the form of figures that look like cross-sections of mountain ranges, with one or more simple or complex peaks (Fig. 6). To get such a figure, it is necessary to collect the liquid dripping out of the column (the "eluate") in a series of consecutively numbered test tubes as separate lots or fractions, and to estimate the protein concentration in each fraction. The concentrations found are then plotted as a graph against the volume of eluate that has already emerged from the column (or against the tube number, which amounts to the same thing). At the places where protein emerged, the curve rises to peaks; where no protein came out, and the liquid coming out of the column was just the salt solution put on the top, the curve hugs the baseline. A single protein should emerge as a single, narrow peak, and a mixture of two proteins, if they are properly separated, gives two distinct peaks.

Clearly, this method of chromatography can be used both to purify proteins and to determine how pure a given preparation is. Complex mixtures of proteins such as blood plasma, egg white or crude aqueous extracts of tissues tend to give complicated patterns with many peaks merging one into another. Preparations of proteins already purified by some means—by using ammonium sulphate or alcohol, for instance (p. 20)—are more likely to give simpler patterns containing only a few components in appreciable amounts; in favourable cases, they may show only a single peak, indicating substantial purity. Pooling the fractions making up any individual sharp peak from a chromatographic run gives a solution containing a presumptively pure protein. If this preparation is run on another column—preferably of a different ion exchanger, and with a different eluting agent—and it again gives a single peak which shows no sign of separating into two, then the presumption is strengthened that only one kind of protein is present.

Hundreds of fractions often have to be collected in a column chromatographic run, and the labour of collecting them by hand would clearly be intolerably tedious (even if laboratory assistants were much easier to get and keep than they are). Fortunately, this is the kind of problem that readily lends itself to automation, and human technicians can be replaced by mechanized ones. Of the equipment one expects to see in a biochemical laboratory, automatic fraction collectors rank among the most inevitable. An instrument of this kind normally has a turntable with spaces for holding the required number of tubes, and some sort of mechanism for moving the turntable round, one space at a time, whenever a given volume of eluate has collected in a tube. A chromatographic run can be left to itself overnight, for instance, and the research worker, on coming into the laboratory in the morning, finds awaiting him a series of tubes containing samples ready for analysis.

What makes ion exchange so important in handling proteins is that other types of chromatography are not generally applicable to molecules of such size and lability. By trying different moving and immobile phases, suitable systems can be found for adsorption or partition chromatography in columns of most groups of compounds, but years of trials failed to produce a kind of system useful for proteins in general (as distinct from a few isolated and more or less atypical proteins). Paper chromatography, too, though adaptable for almost any class of small molecules, proved a disappointment when attempts were made to separate intact proteins with it. Even ion exchange chromatography did not work at first; it was in use for years for separating small molecules like amino-acids (p. 30) before it met with any degree of success among proteins. Special types of ion exchangers had to be developed for handling large molecules, and good ones did not become available until the late nineteen-fifties. Now that they are available, they are a great boon to all who work with proteins.

Handling big molecules (ii) Moving in an electric field—electrophoresis

The principle of electrophoresis is exceedingly simple; it is merely that a charged particle, placed in an electric field, moves towards the electrode of opposite polarity. Proteins carry net charges (at any pH except the iso-electric one) and therefore undergo electrophoresis. Different protein molecules in general carry different amounts of net charge relative to their sizes, and therefore move at different rates, so that they separate, like runners in a race.

The charge on proteins is, of course, sensitively dependent on pH (p. 39), and to get reproducible results it is necessary to fix the pH accurately. Some salt has to be present in the protein solution anyway to make it a good enough electrical conductor, and the salts used are so chosen as to make a "buffer", which fixes the pH and to some extent resists changes of pH when acid or alkali are added.

The apparatus in which electrophoresis can be carried out varies from very simple to quite complex. At its simplest, it is almost as

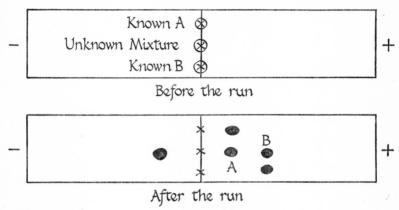

Fig. 7. *Paper electrophoresis*

Before the run, a spot of solution containing unknowns is placed on the starting line, flanked by spots of known compounds to act as markers. After applying a potential difference as shown for a suitable length of time, the paper is dried and sprayed with a reagent to make the spots visible (e.g., ninhydrin for amino-acids). The unknown mixture, in the case illustrated, contained two components presumptively identical with the two markers, together with a third component as yet unidentified. The latter carries a charge of opposite sign to the other two, which may give a hint as to suitable markers to use in a subsequent run.

ludicrously simple as apparatus for paper chromatography. For "paper electrophoresis", all one basically needs is a piece of filter paper, two vessels containing buffer solution, and a source of direct current to two electrodes, one in each buffer vessel. The paper is wetted with buffer solution to make it conducting and the two ends are dipped into the buffer vessels to complete the circuit. The sample, applied before the run as a spot near the middle of the paper, moves one way or the other according to the sign of its electric charge. If a mixture is applied, its components may move at different rates, so that by the end of the run they have formed separate spots (Fig. 7).

Although the final paper looks very much like a finished paper chromatogram (cf., Fig. 2), the roles of paper and liquid in effecting separation are, of course, quite different here. The buffer solution is not a mobile phase but remains more or less stationary during the run, and the paper's only function is to act as an inert supporting medium to hold the buffer. Separation of the spots is caused only by the potential difference applied.

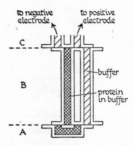

Fig. 8. *Cell for moving boundary electrophoresis*

The glass cell is made of three sections which can be moved sideways with respect to each other so as to bring them in or out of alignment. With all three sections in alignment, the bottom section A is first filled with protein solution. B and C are now moved a little to the right so as to cut their cavities off from A. One limb of the middle section B is filled with protein solution, the other with buffer. C is now moved back to the left to cut it off from B and filled with buffer, leaving the cell as shown. When the run is about to start, B is moved to the left to bring all sections into alignment again and establish sharp protein-buffer boundaries, between A and B on the right and between B and C on the left.

Since electrophoresis here takes place in free solution, with no supporting medium, great care has to be taken to avoid any mixing, which would disturb the boundaries. One source of trouble is that a certain amount of heating due to the passage of current is inevitable, and sets up convection currents which lead to mixing. A simple, ingenious idea which helped to win the Nobel prize for Tiselius was to keep the cell near 4°C, the temperature of the maximum density of water, where temperature change has least effect on the density and convection is therefore minimized.

Paper electrophoresis is very easy to apply effectively to small molecules like amino-acids. The apparatus used in laboratories is often very little more complicated than the basic set-up described above. Applying a potential difference of a few hundred volts, good separations can be achieved in a few hours. To cut down the time required and achieve sharper separations, it is better to apply higher potential differences of several thousand volts, but the paper cannot

BIG MOLECULES—THE PROTEINS

then be allowed to hang free in air; it can be immersed instead in some inert, paraffin-like solvent.

Large molecules like proteins can be run in similar systems with fair success, but difficulties arise because proteins tend to "stick" to the paper. The paper, in other words, is not as inert as it ideally should be, but interacts to some extent with the proteins, often making it difficult to obtain compact spots and clean separations. Many alternatives to paper as the solid medium to support the buffer have been tried. In more complicated kinds of electrophoresis apparatus, however, supporting media are dispensed with altogether, and the whole technique takes a rather different form known as "moving boundary electrophoresis". This type of electrophoresis was first developed into a useful experimental method by Tiselius in Sweden just before the Second World War; Tiselius won the Nobel prize for chemistry in 1948.

As the name suggests, what is here observed is the movement of boundaries rather than of spots. One starts with two boundaries between a solution of protein in buffer and buffer alone. These are set up in a U-shaped glass cell made up of three sections as shown in Fig. 8, the protein solution lying under the buffer. The boundaries are made to move by applying a potential difference; if two or more proteins are present and they move at different rates, two or more boundaries are soon established.

Since proteins are in general colourless, the boundaries cannot be seen without further ado, but ingenious optical systems have been developed for observing them. They give patterns with peaks on them, each peak corresponding to a component of the mixture. The area under each peak is proportional to the concentration of the protein which gives rise to it, so that the technique is useful for quantitatively analysing complex mixtures of proteins such as blood plasma (Fig. 9). It can also be used to determine the purity of a protein, for if only one kind of molecule is present, only one peak appears. For preparing substantial quantities of purified proteins, however, the method is not ideally suited; Fig. 10 and its legend show why this is so.

Although moving boundary electrophoresis is of limited value as a preparative method, it can be used as an analytical tool for following the progress of purification. Suppose, for instance, that a given protein from a living tissue such as liver or potato is to be purified. The tissue might be minced and extracted with cold water or a dilute salt solution; this crude extract is likely to contain many proteins and therefore to show a complex pattern of many peaks on

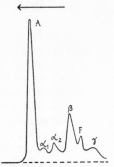

Fig. 9. *Electrophoretic pattern of normal human plasma*
(From Longsworth)

This is the pattern obtained when electrophoresis is carried out in a buffer of pH 8·6, which has been found to give the best separations of plasma proteins for most purposes. The plasma proteins all have iso-electric points below 8·6, so that they carry negative net charges (p. 37) and movement (in the direction of the arrow) is towards the positive electrode.

The pattern shows, by the areas under the respective peaks, the proportions of the various protein components. Changes in these proportions are of some value in the diagnosis of disease. F corresponds to fibrinogen; this peak is not seen in patterns of serum, which is the liquid obtained after blood has clotted. A corresponds to an albumin—a protein with solubility properties rather like the ovalbumin of egg white. The other peaks correspond to globulins—proteins which, unlike albumins, are precipitated by 50% saturation with ammonium sulphate. Many different proteins which happen to move at similar rates contribute to each of the globulin peaks; the number of individual proteins known to be present in plasma is very much greater than the number of peaks that electrophoresis can show. Among the γ-globulins are antibodies (p. 132) responsible for immunity against disease; in some cases, γ-globulin preparations can be given to confer resistance.

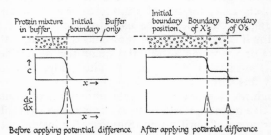

Fig. 10. *Principle of moving boundary electrophoresis*

Although the patterns given by a moving boundary electrophoresis apparatus (Fig. 9) look at first sight rather like those obtained by column chromatography (Fig. 6), they in fact express quite different data. On the horizontal electrophoretic analysis. A preliminary purification step might be fractional precipitation with ammonium sulphate; it might be found that the wanted protein is not in the precipitate formed by 30% saturation with ammonium sulphate, but that it is in the fraction

axis is plotted distance moved, and on the vertical axis not protein concentration, but protein concentration *gradient*—i.e., the rate at which concentration changes with distance along the cell. In the language of mathematics, it is not the curve of concentration (c) against distance (x), but its differentiated version, showing dc/dx against x.

To understand this clearly, consider the situation in one of the limbs of the electrophoresis cell, shown above magnified and turned through 90° as compared to Fig. 8. Before a potential difference is applied, there is a single boundary between the buffer and the protein solution, represented here as containing two protein components which are marked as noughts and crosses. The curve of c against x is flat everywhere except at the initial boundary position; in other words, the rate of change of c with respect to x, dc/dx, is zero everywhere except at this position, which is therefore marked by a single peak in the curve of dc/dx against x. If, during electrophoresis, the noughts move faster than the crosses, the situation soon becomes as shown on the right. There are now two positions at which the protein concentration falls, and two peaks appear in the bottom curve. Since there are more crosses than noughts, there is a greater concentration drop at their boundary, and their peak in the bottom curve is bigger than the peak representing the noughts. It can now be seen how it comes about that the area under a peak is proportional to the concentration of the corresponding component in the mixture. It can also be seen why the method is of limited value for preparative purposes, for to a large extent the noughts and crosses remain unseparated; only the boundaries, not the whole zones occupied by the different components, are quite distinct.

A moving boundary electrophoresis apparatus directly gives the differentiated curve showing dc/dx against x. Its optical system can, it seems, not only measure concentrations but also perform the mathematical technique of differentiation (outdoing in this respect a large proportion even of educated mankind). To explain exactly how it does this is rather complicated, but it is easy to get a rough idea of what is involved.

The principle relied on is that a beam of light is refracted or bent on passing through a region of concentration gradient (because refractive index depends on concentration). Components of the optical system include a cylindrical lens and an inclined slit. A very simple prototype of the system can be made with a test tube, a pencil, some sugar and some water. The tube filled with liquid represents the cylindrical lens, the pencil the inclined slit, and a sugar concentration gradient is set up between a sugar solution and water. The bottom inch or so of the tube is filled with a strong sugar solution, and water is carefully run on top (from a pipette, for instance) so that little mixing takes place. With the tube vertical, the pencil held vertically behind it appears straight. When the pencil is tilted sideways, to correspond to the inclined slit, it acquires a "hump" or peak, as viewed through the tube, at the position of the boundary between the sugar solution and the water.

brought down by 50% saturation. This fraction, freed of ammonium sulphate by dialysis (p. 20), should show a simpler electrophoretic pattern. It might be further purified by ion exchange column chromatography (p. 42). In a favourable case, several discrete peaks might emerge from the column, one of them corresponding to the wanted protein. If the material of this peak shows a single peak in the electrophoresis apparatus too, it is pure by both chromatographic and electrophoretic criteria, giving considerable confidence that the preparation now contains only a single kind of protein.

Handling big molecules (iii) Moving in a centrifugal field— ultracentrifugation

Like electrophoresis, ultracentrifugation sets protein molecules in motion through a solution, at rates which in general differ for different molecules, so that separation is possible. Unlike electrophoresis, however, the rate of motion here is decided primarily not by charge, but by size.

A macroscopic analogy to ultracentrifugation is the simple, well-known mechanical analysis of soils. When a handful of soil is shaken with water and the suspension allowed to stand, the particles settle out in order of size. Gravel falls to the bottom fastest, sand more slowly and clay, finally, slowest of all. For relatively large particles such as these, gravity alone is enough to cause sedimentation. Submicroscopic particles of the order of magnitude of protein molecules, on the other hand, do not fall to the bottom of a solution when it is merely allowed to stand, because the effect of gravity is overwhelmed by random diffusion movements. Only when the effect of gravity is multiplied sufficiently do they sediment.

This is what happens in an ultracentrifuge. Spinning at some 70,000 revolutions per minute, centrifugal forces up to half a million times the force of gravity can be developed. Attaining such speeds and forces naturally raises engineering problems of a different order to those involved in small bench-top centrifuges. An ultra-centrifuge not only needs a more powerful motor to turn it; it also has a pump to evacuate the space in which the rotor spins, so as to minimize air resistance and the consequent heating due to friction. Ultracentrifuges were first developed as useful tools of investigation in Sweden by Svedberg. Like his countryman Tiselius later, he was rewarded by a Nobel prize; his was given in 1926.

During an ultracentrifuge run on a solution of a single protein, a boundary develops between the protein solution at the bottom of the tube and the space above it which has been cleared of protein by sedimentation and now contains solvent only. This boundary, and the rate at which it moves during the run, can be observed by an optical system similar to that used in moving boundary electrophoresis (Fig. 10). If two or more proteins are present and they sediment at different rates, two or more boundaries appear. The method is therefore a good one for determining the purity of a protein preparation. As an additional criterion of purity, it is particularly valuable because it is quite independent of procedures such as ion exchange chromatography and electrophoresis, depending as it does on a different molecular property, the size.

Ultracentrifugation can, of course, be used to follow the progress of protein purification like electrophoresis can, but it has not found much use as a preparative method in itself. It can, however, do one thing which neither ion exchange chromatography nor electrophoresis can do—it can measure molecular weight. The sedimentation rate, given by the observed rate of movement of the boundary, depends on the size and shape of the particles in a way that can be expressed mathematically. With a few relatively simple additional measurements, enough data are obtained to substitute in equations that give the particle weight.

A biological event isolated—the clotting of fibrinogen.
Enzymes as proteins

A little has now been said about the sort of substances that proteins are. But it was emphasized earlier (p. 16) that biochemists are interested not only in what substances are like, but also in what they do—the dynamic aspects being, if anything, more interesting than the static ones. What do proteins do inside living organisms?

As a starting point, let us take the protein fibrinogen, with which we have already struck up a fairly close acquaintance. As is well known, it plays a major role in the clotting of blood, being converted into insoluble fibrin. The agent responsible for effecting the conversion is called thrombin. If, to a solution of fibrinogen, purified as described earlier (p. 21), is added a trace of thrombin, visible changes soon set in. The solution turns turbid and sets to a solid gel, so that the tube it is in can be turned upside down without losing its contents. What has happened is that the molecules of fibrinogen, slightly altered chemically, have aggregated or polymerized to give a continuous network stretching in three dimensions throughout the volume of the solution. When this happens in whole blood, the corpuscles become enmeshed in the network, and so the blood clot is formed.

The thrombin which is the immediate cause of clotting is not, of course, present as such in circulating blood, but is formed from an inactive precursor, prothrombin, by a complex chain of reactions triggered off by tissue damage. Thrombin acts by producing the slight chemical alteration in fibrinogen necessary for it to polymerize. It is an enzyme, and acts catalytically; fibrinogen does not require a stoicheiometric amount of thrombin to convert it into fibrin; a given quantity of thrombin can clot many times its own weight of fibrinogen.

That fibrinogen itself is a protein has already been made abundantly clear. It is even more important to realize that thrombin, too, is a protein—as, indeed, are all enzymes. The significant corollary is that enzymes have the properties of proteins. Some of the most characteristic properties of enzymes are, indeed, directly attributable to their protein nature. Most striking among these properties are, firstly, their susceptibility to treatments which denature proteins, notably heating; and, secondly, the way in which their activity is closely dependent on pH, changing markedly even with pH changes too small to cause denaturation.

Temperature sensitivity is easy enough to show. If a little thrombin solution in a test tube is heated to 100°C by putting it into a boiling water bath for a few minutes, it is found on cooling to have lost the power to clot fibrinogen. Enzymes in general behave in this way. In some quite narrow range of temperature, usually well below the boiling point of water, the rate at which denaturation occurs increases quite suddenly, so that by the time 100°C is reached it appears to be instantaneous. Extremes of pH, too, cause denaturation and destroy enzyme activity; few enzymes can withstand one-tenth normal acid or alkali (pH 1 or 13 respectively), and many are denatured even by milder pH conditions.

Much smaller changes of pH also affect the activity of enzymes, though in a different way. Thrombin, for instance, clots fibrinogen only if the pH of the reaction mixture is within the range of approximately 6 to 10. This limitation is due not to denaturation, but to changes in the state of ionization; it is not a question of destroying the unique arrangement in which the peptide chains are folded, but only of changes in charged groups. As has been said, the ionization ranges (i.e., the regions near the pK's) of the various kinds of ionizing groups in proteins cover the whole range of pH from quite strongly acid to quite strongly alkaline (p. 39 and Table II). Anywhere between the two extremes, therefore, a change in pH must have an effect; a rise in pH adds negative charges, takes away

BIG MOLECULES—THE PROTEINS

positive ones, or both, while a fall has the opposite effects. For an enzyme to be active, it needs a particular pattern of charged groups just as much as it needs proper folding of its chains. At the particular pH at which the right ionic species makes the greatest contribution, the activity of an enzyme is at a maximum; this is its optimum pH. On both sides of the optimum, the activity falls off, usually quite rapidly. Hence, in working with enzymes, it is important to fix the pH accurately by means of appropriate buffers (p. 45). It is comparatively rare for the activity of an enzyme to remain appreciable more than two or three pH units away from its optimum.

The effect which deprives the enzyme of its activity is in this case freely and instantaneously reversible, being only a matter of ionic equilibria. It is only necessary to adjust the pH back to near the optimum for activity to show itself again. This pH effect thus differs from denaturation by heating or more drastic changes of pH, which is in general irreversible; an egg that has been cooked by boiling cannot be uncooked by cooling.

Once it is realized that important properties of enzymes reflect their protein nature, the preoccupation of biochemists with proteins is a mystery no longer—for enzymes are, after all, the very stuff of which biochemistry is made. Enzymes are the agents that make things happen in living organisms; and, though not all proteins are enzymes, all enzymes are proteins. Of the protein inside cells, making up its "protoplasm", a large proportion is enzyme protein. The reactions these enzymes catalyse make up in total the metabolism which supports—in fact, which *is*—the complex process of maintenance and growth of which life basically consists.

A single enzyme reaction proceeding in a test tube, therefore, is a fragment of the dynamics of life—a biological event isolated. It is a little but important piece of living function, just as a protein molecule is a little but important piece of living form. In this lies its significance and its fascination.

Considered as a particular illustration of this wide category of enzyme-catalysed reactions, the clotting of fibrinogen by thrombin stands out for the vividness with which it strikes the naked eye. (It is not in all ways typical—especially in that both the enzyme and the substrate are in this case proteins; in other cases—the majority—only the enzyme is a protein, and the substrate on which it acts belongs to some other class of compounds.) Any enzyme reaction can be followed by suitable chemical means, but in most other cases there is no change that is directly visible without further ado.

The clotting of fibrinogen by thrombin thus gives a particularly striking epitome of the biochemical approach to living phenomena. Here, one can actually see biological molecules, in purified form, performing an *in vitro* version of an important *in vivo* process. One might go so far as to say that this reaction makes a good test for finding out whether a student has the makings of a biochemist in him. Does he get any feeling of satisfaction out of seeing two clear, colourless solutions, when mixed, carry out the prototype of a biological process? Does he get an emotional kick out of the knowledge that, going on in the test tube before his eyes, is a biological event isolated? If he does, then he is a biochemist at heart.

III

Life's Basic Device—Specific Catalysis by Enzymes. The Anaerobic Breakdown of Sugar

A worldly motive—the thirst for alcohol

Is Noah to be considered the world's first biochemist? His acquaintance with alcoholic fermentation is, after all, described in the Book of Genesis—and the study of fermentation came to play a decisive part in the emergence of modern biochemistry. From it, more than from any other single area of enquiry, arose the idea of the ubiquitous role of enzymes—the idea that specific catalysis by a variety of enzymes is the means by which metabolism proceeds, and hence may well be regarded as life's basic device. This concept is probably the most fundamental that biochemistry has formed.

Man's interest in alcoholic fermentation needs no explanation in itself. Understandably, stimulus for scientific discovery has often come from aims less exalted than the disinterested pursuit of knowledge. Many of the alchemists who prepared the way for modern chemistry were motivated by the greed for gold. Biochemistry owes at least as much to man's thirst for alcohol as chemistry does to his appetite for money. Early impetus came from the worldly quest for liquor in the one case, for lucre in the other.

The story of fermentation illustrates particularly well the origin of modern biochemistry as a whole from classical biology and chemistry. Biochemistry as we know it arose less as a harmonious fusion of the two other sciences than as the outcome of a controversy between them that was bitter at times. In the nineteenth century there was a major tug-of-war between the chemical and biological viewpoints. Each claimed the study of fermentation as its own, and in the context of contemporary thought it was difficult for it to belong to both. The resolution of the conflict, naturally, came in terms that are characteristically biochemical. It hardly seems too fanciful to see a process of Hegelian dialectic at work in the development that took place. The thesis was chemical, the antithesis biological; the synthesis that resulted was biochemical.

Chemical thesis—the overall result

At first, of course, the chemistry *versus* biology issue did not arise. Alcoholic fermentation was practised successfully even in prehistoric times without any idea that its cause lies with living organisms. It was possible to state the overall chemical outcome of the process correctly without worrying about the nature of the causative agent. This was, in fact, done by Lavoisier at the end of the eighteenth century. It was one of the very earliest applications of his new theory of chemistry, made within a few years of the momentous advance of learning to think in terms of elements such as carbon, hydrogen and oxygen. Unlike the "elements" that had been proposed previously (p. 130), these remain constant in amount, and it is therefore possible to set up balance sheets in terms of them. Lavoisier was able to show that essentially all the carbon, hydrogen and oxygen of the sugar broken down reappear in the form of alcohol and carbon dioxide. In modern formulae,

$$C_6H_{12}O_6 = 2C_2H_5OH + 2CO_2$$

(Actually, Lavoisier also found small quantities of acetic acid. His results were altogether more clear-cut than they should have been, since there were major errors of up to 50% in his analyses, and the final agreement was good only because of the fortunate way in which the errors cancelled each other. Genius is, perhaps, not so much the infinite capacity for taking pains as the knack of having luck consistently.)

Lavoisier did not particularly concern himself with the agent that brings fermentation about. Yeast cells had been observed microscopically by Leeuwenhoek, that industrious Dutch grinder of lenses, and described by him in his letters to the Royal Society in London as early as 1680, but his reports were little heeded. In the early nineteenth century Berzelius, the Swedish autocrat of the chemical world, introduced the notion and the name of catalysis, and thought of yeast as a catalyst for the conversion of sugar into carbon dioxide and alcohol. He saw no reason to regard it as anything more than a rather remarkable chemical substance.

Biological antithesis—yeast as a living organism

The world of science did not take easily to the fact that yeast is, in fact, a living organism, though the claim was made in 1837 by no less than three independent workers (Cagniard-Latour, Schwann and Kützing). Berzelius at first scornfully dismissed the evidence;

in Germany, Liebig—a pioneer in applying the new chemistry to biological problems—saw fit to publish in his *Annalen* an elaborate take-off ridiculing the sort of microscopic observations on which the claim was based. The anonymous author of this piece of scientific humour was none other than Liebig's close friend Wöhler, whose synthesis of urea has become so famous (p. 69).

Modern scientific journals rarely offer such spicy reading as the *Annalen* for 1839. Controversies nowadays tend to be genteelly hedged about with polite restraint. Not so in those lusty days—Wöhler laid about him with heavy Teutonic sarcasm. He began by picturing the personal discomforts of a long, cold night spent at the brewery in following the life-history of the microscopic organisms, whose fine structure he then described in ludicrously fanciful detail:-

"It is possible to distinguish clearly a stomach and intestine, the anus (a pink spot) and the urine-forming organs. From the moment they escape from the egg, these animals visibly gulp down sugar out of the solution, and one can see it quite clearly arriving at the stomach. There it is instantaneously digested, as shown by the expulsion of excrement which follows promptly. In short, these infusoria feed on sugar and release alcohol from the bowels and carbonic acid from the urinary organs".

With so much weight behind the opposition to the view that the causative agent of fermentation is a living organism, the controversy became a *cause célèbre* of the mid-nineteenth century. It came to be associated principally with the names of Liebig on the one hand and Pasteur on the other—becoming thus a battle between two scientific giants of the century.

Liebig's view was that the "ferment" is an extraordinarily labile organic substance which is formed by interaction of air with something in plant juices and which effects breakdown by transferring its own remarkable instability to sugar. Pasteur, however, had satisfied himself by 1860 that yeast can grow, with increase in dry weight, in a simple medium containing only sugar and salts, showing that the fermenting agent is something more than decaying plant matter. It was clear to him that fermentation is brought about by a living organism itself capable of multiplication and growth, not by a plant substance in a state of communicable decomposition. The formerly great Liebig, grown stubborn with age, fought a pathetic rearguard action, but even he remained silent after Pasteur had driven his victory home in 1871 with a taunting challenge:—he "offered to prepare in a mineral medium, in the presence of a commission to be chosen for the purpose, as great a weight of ferment as Liebig could reasonably demand". (Although the controversy was not one between France and Germany, both views finding adherents in both countries, it does seem to have been intensified by overtones of national rivalry, in an age when patriotism seemed the noblest of ideals. Pasteur was concerned to salvage French prestige after military disaster. "Our misfortunes inspired me with the idea of these researches" he wrote in the preface of a book describing his later experiments on fermentation. "I undertook them immediately after the war of 1870, and have since continued them without interruption, with the determination of perfecting them, and thereby benefiting a branch of industry [brewing] wherein we are undoubtedly surpassed by Germany." Seeking in the brewery the *gloire* lost on the battlefield seems a strange motive to have led to epoch-making scientific work.)

Biochemical synthesis—a multi-enzyme system

Fermentation was now firmly established as a "physiological act", correlated with the life and organization of yeast cells. "No fermentation without life" was Pasteur's contention, and the evidence for it was impressive.

Yet he had been dead only two years when, in 1897, Eduard Buchner of Tübingen managed to prepare a cell-free extract of yeast capable of fermenting sugar—an extract which, though derived from living organisms, could not be said to contain any. To adherents of Pasteur's views, this seemed as remarkable, perhaps, as it would seem to-day to hear of a cell-free extract of brain capable of thinking. A process that they had come to regard as inseparably linked to the activity of intact living organisms could now be brought about by a mere mixture of molecules floating about in solution.

Like many other discoveries, this one arose from a chance observation during work aimed at something quite different. Eduard Buchner, his brother Hans and their assistant Hahn were trying to prepare an extract of yeast for therapeutic purposes. To do this, they ground yeast with sand to break the cells, added kieselguhr and squeezed the liquid out of the mixture in a press. The problem now arose of how to stop the fluid from going bad. Most of the ordinary antiseptics were excluded, being too poisonous for human consumption, and the Buchners therefore resorted to "preserving" in the kitchen sense—adding sugar, that is, as in making jam, to give a concentration high enough to prevent the growth of micro-organisms. To their surprise, they noted all the signs of vigorous fermentation in the resulting solution. (Had they tested its "therapeutic" action on a patient, the effect would presumably have been equally surprising, though brief.)

The Buchners were thus able to conclude "that the production of alcoholic fermentation does not require so complex an apparatus as the yeast cell, and that the fermenting power of yeast juice is due to the presence of a dissolved substance." This juice, an opalescent, brownish-yellow fluid, became something of an obsession with biochemists for close on half a century. Unravelling, piece by piece, the events that go on in it was a hitherto unparalleled triumph for them, and the facts that emerged still make up one of the most significant areas of biochemical knowledge. The "dissolved substance" of the Buchners, to which the name zymase was originally given, has turned out to be a mixture of substances—a whole battery of enzymes, and several non-protein substances or coenzymes necessary for some of the enzymes to function. The route by which glucose breaks down to alcohol has been fully mapped out and found to involve about a dozen steps, each of which is catalysed by a different, specific enzyme present in yeast juice.

Before leaving the historical aspect of alcoholic fermentation, it is worth reflecting on the paradoxical light the newer work sheds in retrospect on Liebig's position in his controversy with Pasteur. Liebig was clearly wrong in the context of the discussion in which he was engaged—yet his view contained the germ of an even more fundamental truth. He thought as a chemist, and felt that the problem of fermentation should be tackled chemically; hence his reluctance to see it pass into the region of life processes, which many at that time felt to be beyond the reach of detailed chemical attack. The chemical transformations occurring in living cells seemed so unapproachable that a biological explanation virtually amounted to a renunciation of chemical interest (p. 70). Pasteur, indeed, having established that fermentation is linked to the life of yeast, freely confessed complete ignorance of its more detailed chemical nature and cause. Though he had started his career as a chemist, his famous researches on micro-organisms had given him much more the outlook of a biologist; hence he was readier to be satisfied with the explanation of fermentation as a "physiological act".

Only with the Buchners' yeast juice did it become possible for the two viewpoints to interact fruitfully and lead to the characteristically biochemical conclusion that fermentation is caused by a multienzyme system.

An unexpected participant—involvement of phosphate

The stages of alcoholic fermentation are summarized in Fig. 11. A first impression on glancing at them may well be surprise that most of the intermediates carry phosphate groups. This is a fact of the greatest significance, which could never have been guessed from the overall equation for fermentation, in which phosphate does not appear.

What led to its discovery was an observation made by Harden and Young in London in 1905. They were investigating why fermentation by yeast juice (as distinct from fermentation by intact, living yeast) does not continue very long. When sugar is added, the solution at first bubbles and froths merrily, but gradually the evolution of carbon dioxide slows down and stops, even though plenty of sugar is left. Harden and Young found that if, at this stage, an inorganic phosphate is added, a fresh burst of activity is called forth. During fermentation, the inorganic phosphate disappears from the solution and is turned into organic phosphate esters. Eventually, fermentation grinds to a halt for lack of inorganic phosphate. One phosphate ester that accumulates is fructose-1,6-diphosphate; it was isolated by Harden and Young and known for some time by their names. This work was the beginning of the meteoric rise of phosphate to biochemical fame, and Harden was awarded a half-share of the Nobel prize for chemistry in 1929.

Of all the many phosphate compounds now known to occur in cells, perhaps the most important is adenosine triphosphate, or ATP for short. About the chemical structure of this compound (cf. p. 176), it is enough for the moment to know that the three phosphate groups are attached in a row to the adenosine moiety:—

$$\text{adenosine}-\text{O}-\overset{\overset{\displaystyle O}{\|}}{\underset{\underset{\displaystyle OH}{|}}{P}}-\text{O}-\overset{\overset{\displaystyle O}{\|}}{\underset{\underset{\displaystyle OH}{|}}{P}}-\text{O}-\overset{\overset{\displaystyle O}{\|}}{\underset{\underset{\displaystyle OH}{|}}{P}}-\text{OH}$$

If Ⓟ is used as shorthand for a phosphate group, this can be more compactly written A—Ⓟ—Ⓟ—Ⓟ.

ATP is a very vital substance for living organisms to have in constant supply. It is used for a variety of purposes (Chap. VIII), being itself broken down in the process, usually to inorganic phosphate, HOⓅ, and the diphosphate, ADP. A large part of the activity of cells is directed towards resynthesizing ATP from ADP and inorganic phosphate. This very process is, indeed, the main *raison*

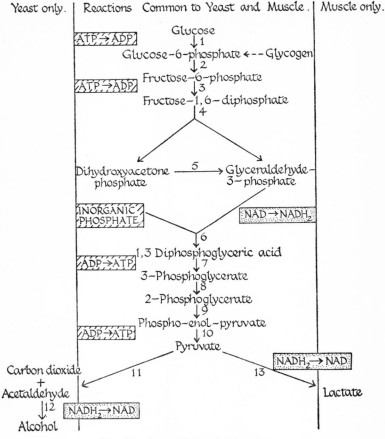

Fig. 11. *Anaerobic breakdown of sugar*

Each numbered step is catalysed by a specific enzyme. All of these enzymes have been purified to a greater or lesser extent, and some have been crystallized. They are given systematic names which denote the substrate or substrates acted on and the type of reaction catalysed; many of them have, in addition, a less cumbersome trivial name. All the names end in "-ase", a suffix which is reserved for and characteristic of enzymes (except in a very few cases of old, well-established names such as thrombin). For instance, the enzyme for step 11 is called pyruvate decarboxylase.

In the first step, the terminal phosphate group of ATP (see text) is transferred to position 6 of glucose, leaving ADP and glucose-6-phosphate.

THE BIOCHEMICAL APPROACH TO LIFE

```
1    CHO                CHO
     |                  |
2    HCOH               HCOH
     |                  |
3    HOCH               HOCH
     |          1       |
4    HCOH  + ATP  ──→   HCOH    + ADP
     |                  |
5    HCOH               HCOH
     |                  |
6    CH₂OH              CH₂O-Ⓟ
```

(It may look at first as though a hydrogen atom disappears in this reaction, but this is not the case, as can be seen by writing the formulae more fully. To avoid further complications, the formulae given are those of the undissociated phosphoric acids rather than the phosphate ions which occur physiologically.

$$\begin{array}{l} \text{CHO} \\ | \\ \text{HCOH} \\ | \\ \text{HOCH} \\ | \\ \text{HCOH} \\ | \\ \text{HCOH} \\ | \\ \text{CH}_2\text{OH} \end{array} \; + \; \text{adenosine—O—}\overset{\overset{O}{\|}}{\underset{\underset{OH}{|}}{P}}\text{—O—}\overset{\overset{O}{\|}}{\underset{\underset{OH}{|}}{P}}\text{—O—}\overset{\overset{O}{\|}}{\underset{\underset{OH}{|}}{P}}\text{—OH}$$

$$\Big\downarrow 1$$

$$\begin{array}{l} \text{CHO} \\ | \\ \text{HCOH} \\ | \\ \text{HOCH} \\ | \\ \text{HCOH} \\ | \\ \text{HCOH} \quad \text{O} \\ | \qquad \quad \| \\ \text{CH}_2\text{O—P—OH} \\ \qquad \quad | \\ \qquad \quad \text{OH} \end{array} \; + \; \text{adenosine—O—}\overset{\overset{O}{\|}}{\underset{\underset{OH}{|}}{P}}\text{—O—}\overset{\overset{O}{\|}}{\underset{\underset{OH}{|}}{P}}\text{—OH}$$

The misleading impression results from a shortcoming in the otherwise very useful convention of writing Ⓟ as abbreviation for a phosphate residue.)

Step 2 is an internal rearrangement or isomerization, involving no change in the number of atoms in the molecule, but only the shift of two hydrogen atoms from position 2 to position 1.

```
     CHO                CH₂OH
     |                  |
     HCOH               CO
     |                  |
     HOCH               HOCH
     |           2      |
     HCOH       ──→     HCOH
     |                  |
     HCOH               HCOH
     |                  |
     CH₂O-Ⓟ             CH₂O-Ⓟ
```

LIFE'S BASIC DEVICE—SPECIFIC CATALYSIS BY ENZYMES

The newly formed $-CH_2OH$ group at position 1 is phosphorylated at the expense of a second molecule of ATP (step 3); this reaction resembles that of step 1. Next follows a split into two three-carbon fragments.

$$\begin{array}{c} CH_2O\!\!\!\!\!\!\text{\textcircled{P}} \\ | \\ CO \\ | \\ HOCH \\ \text{-----} \\ HCO{\,}H \\ | \\ HCOH \\ | \\ CH_2O\!\!\!\!\!\!\text{\textcircled{P}} \end{array} \xrightarrow{4} \begin{array}{c} CH_2O\!\!\!\!\!\!\text{\textcircled{P}} \\ | \\ CO \\ | \\ CH_2OH \\ \text{Dihydroxyacetone} \\ \text{phosphate} \end{array} + \begin{array}{c} CHO \\ | \\ HCOH \\ | \\ CH_2O\!\!\!\!\!\!\text{\textcircled{P}} \\ \text{Glyceraldehyde-} \\ \text{3-phosphate} \end{array}$$

Dihydroxyacetone phosphate is an isomer of glyceraldehyde phosphate, and step 5 merely involves another rearrangement, the shifting of two hydrogen atoms from a terminal to the middle carbon atom.

Reaction 6 is the peculiar but significant one involving the pick-up of inorganic phosphate, which is condensed with the carboxyl group arising by oxidation of the aldehyde at the expense of NAD (see p. 65). The significance of this reaction is more fully discussed later (p. 149).

$$\begin{array}{c} CHO \\ | \\ HCOH \\ | \\ CH_2O\!\!\!\!\!\!\text{\textcircled{P}} \end{array} + HO\!\!\!\!\!\!\text{\textcircled{P}} + NAD \xrightarrow{6} \begin{array}{c} COO\!\!\!\!\!\!\text{\textcircled{P}} \\ | \\ HCOH \\ | \\ CH_2O\!\!\!\!\!\!\text{\textcircled{P}} \end{array} + NADH_2$$

The carboxyl-bound phosphate is then transferred to ADP to give ATP.

$$\begin{array}{c} COO\!\!\!\!\!\!\text{\textcircled{P}} \\ | \\ HCOH \\ | \\ CH_2O\!\!\!\!\!\!\text{\textcircled{P}} \end{array} + ADP \xrightarrow{7} \begin{array}{c} COO^- \\ | \\ HCOH \\ | \\ CH_2O\!\!\!\!\!\!\text{\textcircled{P}} \end{array} + ATP$$

Step 8 is yet another isomerization, but this time it is the phosphate group which changes position, shifting from position 3 to position 2. From the product, the elements of water are removed.

$$\begin{array}{c} COO^- \\ | \\ HCO\!\!\!\!\!\!\text{\textcircled{P}} \\ | \\ CH_2OH \end{array} \xrightarrow{9} \begin{array}{c} COO^- \\ | \\ CO\!\!\!\!\!\!\text{\textcircled{P}} \\ \| \\ CH_2 \end{array} + H_2O$$

The extraordinary-looking compound which results is the phosphorylated derivative of the enol form of pyruvate. Accordingly, when its phosphate group is transferred to ADP to form more ATP (step 10) pyruvate is left.

$$\begin{array}{c} COO^- \\ | \\ COH \\ \| \\ CH_2 \\ \text{pyruvate (enol form)} \end{array} \qquad \begin{array}{c} COO^- \\ | \\ C\!=\!O \\ | \\ CH_3 \\ \text{pyruvate (keto form)} \end{array}$$

d'être of the machinery of fermentation. Yeast cells are not interested in alcohol for its own sake (differing in this respect from some other living organisms); the alcohol is merely an end-product that the process happens to give, on a par with carbon dioxide. To maintain a constant supply of ATP, on the other hand, is essential for life and growth.

Looking at the steps of fermentation with this in view, it is a little disconcerting to find that, in the very first step, a molecule of ATP is used up, donating its terminal phosphate group to glucose. The deficit on the ATP balance sheet is further increased to two by step 3, in which a phosphate group from another ATP molecule is transferred to the sugar phosphate to give a diphosphate. These two ATP molecules used in the early stages can, however, be regarded as a deposit, to be returned later with interest, for in steps 7 and 10 phosphate groups are transferred to ADP to give ATP. At first sight, it may look as though this only just balances the account, but in fact it produces a surplus of two ATP molecules. This is because two molecules of glyceraldehyde phosphate are formed from each molecule of glucose; fructose-1,6-diphosphate splits to give one molecule of it directly and another one indirectly via dihydroxyacetone phosphate and the rearrangement of step 5. All the steps from 6 to 11 involve three-carbon substances, and therefore occur twice during the breakdown of each molecule of the six-carbon sugar, so that four molecules of ATP are synthesized by steps 7 and 10.

The net gain of two molecules of ATP is the aim and object of the whole complicated exercise. The extra phosphate required derives from the inorganic phosphate picked up in step 6. Here, then, is the explanation of the disappearance of inorganic phosphate noted by Harden and Young. (In active, intact yeast, as distinct from yeast juice, ATP is split as it is used to give ADP and inorganic phosphate which keep the whole reaction series going.)

Getting along without oxygen—oxido-reduction aspect

As the overall equation shows, fermentation is an anaerobic process. "*La vie sans l'air*", or "life without oxygen", was Pasteur's

---

Notes to Fig. 11 continued.

The loss of carbon dioxide from pyruvate (step 11) leaves acetaldehyde, which is reduced by $NADH_2$ in the final step.

$$CH_3COCOO^- + H^+ \xrightarrow{11} CH_3CHO + CO_2$$
$$CH_3CHO + NADH_2 \xrightarrow{12} CH_3CH_2OH + NAD$$

definition of it. Nevertheless, individual steps do involve oxidation and reduction. In step 6, the aldehyde group of glyceraldehyde phosphate is oxidized to the level of a carboxyl group, while in step 12 acetaldehyde is reduced to alcohol.

These oxidations and reductions involve a hydrogen carrier known as NAD, the abbreviation for nicotinamide adenine dinucleotide (p. 125). (This name has only recently been introduced; earlier names used in many books for the same compound are DPN, standing for diphosphopyridine nucleotide, and coenzyme I or CoI). NAD readily undergoes reversible reduction, accepting and giving up again a pair of hydrogen atoms.

$$NAD \underset{-2H}{\overset{+2H}{\rightleftarrows}} NADH_2$$

In step 6, NAD is reduced; in step 12 it is reoxidized. The quantity of NAD reduced is balanced by the amount reoxidized. Cells, therefore, need provide themselves with only small, catalytic amounts of NAD, since each molecule can be used over and over again. The enzymes that catalyse steps 6 and 12 cannot work without NAD (or $NADH_2$, as the case may be), and it is for this reason that NAD is referred to as their "coenzyme".

Another yeast enzyme which collaborates with NAD catalyses a reaction in which dihydroxyacetone phosphate is reduced to glycerol phosphate.

$$\begin{array}{c} CH_2O\circled{P} \\ | \\ CO \\ | \\ CH_2OH \end{array} + NADH_2 \rightarrow \begin{array}{c} CH_2O\circled{P} \\ | \\ CHOH \\ | \\ CH_2OH \end{array} + NAD$$

This reaction is not shown in Fig. 11, since it does not lie on the direct route leading from glucose to alcohol. It does, however, always occur to some extent when yeast ferments sugar, and accounts for the formation of small quantities of glycerol, which arises by hydrolytic removal of the phosphate ester group. It is largely the glycerol in wines which gives them their "body".

Normally, most of the $NADH_2$ produced by reaction 6 is snapped up for the reduction of acetaldehyde by reaction 12. Little $NADH_2$ being available for reducing dihydroxyacetone phosphate, much less glycerol than alcohol is formed. If, however, reaction 12 is interfered with, the yield of glycerol is increased.

Effective ways of doing this were sought in Germany during the war of 1914–1918, when glycerol was much in demand for making explosives. One simple method is to add sodium bisulphite to the fermenting liquor; this traps the acetaldehyde as it is produced, forming the acetaldehyde-bisulphite compound well known in organic chemistry and thereby preventing reaction 12 from taking place. Extra glycerol therefore replaces alcohol among the fermentation products. Like the choice between guns and butter, yeast offers either glycerol or alcohol as war or peace alternatives.

The biochemical unity of living matter—the similarity of yeast and muscle

The significance of the work on yeast juice soon turned out to be wider than had ever been dreamed, because it was found that extracts of higher animals are also capable of catalysing many of the reactions performed by yeast juice. Mammalian muscle, a tissue particularly active in breaking down carbohydrate, was used most, and work on extracts of it went on largely in parallel to that on yeast juice during the first half of the present century.

Fig. 11 shows how great is the area of agreement between yeast and mammalian muscle about the route of sugar breakdown. It is a remarkable demonstration of the biochemical unity of all living matter that cells so widely apart in biological classification should differ so slightly in a major metabolic pathway. This unity contrasts strangely with the fascinating variety which could not fail to impress classical biologists studying whole plants and animals. The further living organisms are taken to pieces, the more alike they become; as the cell theory began, so biochemistry continues to emphasize basic similarities of ground-plan (cf. p. 140). The molecular differences which underlie the macroscopic diversity of living forms are so small, indeed, that biochemists have as yet devoted relatively little attention to them (cf., however, Fig. 3). So far, they have concentrated mainly on establishing the broad features of the chemistry of life, with respect to which an impressive degree of uniformity has come to light.

Comparing carbohydrate breakdown in yeast and muscle, it is found that the main differences lie at the beginning and the end of the pathways. At the beginning, the major starting material for muscle is glycogen, a polymeric form of glucose, rather than the free sugar. Ultimately, however, the muscle's reserve of glycogen is formed from glucose picked up from the blood, and its formation involves the first step of Fig. 11, so that this difference does not go

LIFE'S BASIC DEVICE—SPECIFIC CATALYSIS BY ENZYMES

deep. (Yeast also forms a certain amount of glycogen, anyway.)

Near the end lies a more significant divergence of the pathways, for in muscle pyruvate does not give acetaldehyde. Instead of being decarboxylated before reduction, it is reduced directly under the influence of another specific enzyme, the product being lactate. The two equivalents of hydrogen for the reduction again come from $NADH_2$, so that in its oxido-reduction aspect, lactate formation in muscle is like alcohol formation in yeast, reaction 13 replacing reaction 12 as the mechanism for reoxidizing $NADH_2$ produced in reaction 6.

$$\begin{array}{c} COO^- \\ | \\ CO \\ | \\ CH_3 \end{array} + NADH_2 \xrightarrow{13} \begin{array}{c} COO^- \\ | \\ CHOH \\ | \\ CH_3 \end{array} + NAD$$

Muscle can, of course, not only break down glycogen as far as lactate; it can also degrade it completely to carbon dioxide and water. The latter process is much superior in that it provides far more ATP (p. 150). It does, however, require oxygen.

$$(C_6H_{10}O_5)_n + 6nO_2 \rightarrow 6nCO_2 + 5nH_2O$$

Although the muscle carries its own store of glycogen and can therefore remain independent of blood glucose for some time, it can store only small quantities of oxygen. During heavy work, oxygen brought by the blood soon comes to be in short supply, and some of the glycogen is broken down only as far as lactate. This accounts for the well-known fact that the lactate concentration in the blood rises if exertion is at all prolonged.

Fig. 11 shows how slight is the twist of biochemical fate which determines that it should be lactate and not alcohol that is poured into the blood under stress. Were the enzyme make-up of human muscles just that little bit different, hard physical exercise might be more popular than it is—the opposite of a sobering thought.

The role of enzymes in biological thought—nineteenth century "vitalism"

Having seen a little of what grew out of it makes it possible to appreciate more fully the significance of the Buchners' discovery of fermentation by yeast juice. In a nutshell, it opened the way to the work which has shown the feasibility of tracing the pathways of intracellular metabolism. That this can be done has not always seemed as obvious as it does nowadays.

Among the ill-defined kinds of nineteenth century thought loosely termed "vitalism", there were two strands in particular which deterred scientists from tackling the chemistry of life. As a convenient over-simplification, it could be said that it was Wöhler's synthetic urea and the Buchners' fermenting yeast juice respectively which made these two deterrents seem less forbiddingly ultimate than before.

The first deterrent was the view that "in living nature, the elements seem to obey laws quite different from those in inorganic nature". These are the words of the great Berzelius's own statement of his beliefs, written early in the nineteenth century. "If one could discover the cause of this difference," he went on, "one would have the clue to the theory of organic [i.e., biological] chemistry. But this theory is so abstruse that we have no hope of discovering it, at least at present.... The essence of the living body lies not in its elements, but in some other principle. This principle, which we call the *vital* or *assimilating force*, is not inherent in the elements, and does not constitute one of its primordial properties like weight, impenetrability, electric polarity etc.; we cannot conceive what it does consist of, how it originates or how it ends."

To appreciate what Berzelius had in mind when he wrote this, it must be remembered that the chemical thought of his time was dominated by the dualistic theory, according to which combination takes place between elements or groups of opposite charge. Although this principle met with great success in systematizing facts about salt-like inorganic compounds, it required a degree of intellectual contortionism to see it at work in complex carbon compounds. No wonder Berzelius thought that "in living nature, the elements seem to obey laws quite different from those in inorganic nature". Inorganic compounds could be made artificially in the laboratory, but organic ones seemed to be the products of an agency quite different from the ordinary, inanimate forces of physics and chemistry.

A vital force whose sole *raison d'être* it was to be the architect of carbon compounds, however, clearly could not survive the great successes of synthetic organic chemistry as the nineteenth century drew on. It would be a misleading over-dramatization to say that belief in such a vital force shattered once for all on the hard, white crystals of Wöhler's synthetic urea in 1828. As Wöhler himself pointed out, his starting materials included organic ones. His

LIFE'S BASIC DEVICE—SPECIFIC CATALYSIS BY ENZYMES

synthetic urea—identical with that of a specimen "made in every respect by himself"—was formed from ammonium cyanate.

$$NH_4CNO \rightarrow CO(NH_2)_2$$

On February 22nd, 1828, he wrote as follows to Berzelius, under whom he had studied and to whom he remained devoted:—

"It is noteworthy that for the production of cyanic acid (and also of ammonia) one still has to start in the first place from organic matter. A philosopher would say that the organic has not completely disappeared from the animal charcoal and the cyanates formed from it, and that it is for this reason that it is still possible to prepare an organic substance from them".

Cyanates were prepared at this time by heating organic matter such as dried blood, hoofs or horns with iron and potash; the resulting potassium ferrocyanide yielded potassium cyanide on heating, and the cyanate was obtained from this by oxidation. Among the hoofs, the shrewd observer can still spot the cloven one of "vitalism".

In 1845, however, Kolbe achieved the first true total synthesis by making acetic acid from its elements. Other syntheses followed, and gradually the mystique fell away from carbon compounds as such. Organic synthesis ceased to be the exclusive prerogative of living systems; the term "organic chemistry" acquired its modern meaning and no longer denoted a connection with living organisms, as it had originally. By 1860, it was possible for Kekulé, in his classical text-book, to define organic chemistry as "the chemistry of carbon compounds", and to go on to emphasize that "organic chemistry does not deal with the study of the chemical processes in the organs of plants or animals. This study forms the basis of physiological chemistry".

The second deterrent was concerned with precisely these processes, as distinct from the structures of the substances which undergo them. It took the form of scepticism about the feasibility of tracing the chemical reactions undergone by a substance once it has entered a living cell. To use Berzelius's own words again, it was the view that "the highest knowledge which we can attain is the knowledge of the nature of the productions, whilst we for ever are excluded from the possibility of explaining how they are produced." What goes in and what comes out can be studied chemically, but what goes on in between, it seemed, remains hidden from the chemist's prying eyes.

Pessimism on this point was widespread later in the nineteenth century. The idea had got abroad of "protoplasm" as a sort of vague, superchemical entity of unanalysable complexity. Chemists tended to feel that, having watched their molecules disappear into the jaws of this monster, there was nothing to do but sit back and wait until the end-products came out.

Success in mapping the route of sugar breakdown by yeast and animal extracts changed all that. The intermediates of carbohydrate metabolism turned out to be identifiable molecules of manageable size. Their conversions, one into another, were explicable by catalysis at the hands of specific enzymes, whose action could be studied in isolation in the test tube. If alcoholic fermentation—Pasteur's "physiological act"—is reducible to a consecutive series of enzyme-catalysed reactions, why should the same not be possible for other metabolic processes?

Enzymes thus came to appear as the keys to living nature. It is hardly possible to over-emphasize how completely they have come to permeate biochemical thought. If one had to pick out one single notion as being the most basic to biochemistry, it should probably be that of specific catalysis by enzymes. Natural sciences, a philosopher (Broad) has written, "flounder about in the dark till some man of genius sees what are the really fundamental factors and the really fundamental structure of the region of phenomena under investigation. In mechanics, the keystone is the notion of acceleration; in chemistry it is the theory of elements and compounds and the conservation of mass..." . In biochemistry, one might add, it is the idea of catalysis by enzymes that occupies the crucial position.

Enzymes were not unknown before the work on yeast juice, but there was little reason to suspect the full extent and importance of their role in metabolism. The earliest enzymes to be studied were mostly of the kind which are secreted and act hydrolytically outside cells to digest foodstuffs—the digestive enzymes of the alimentary tract, for instance. From these, it was a far cry to the idea that there might be a multitude of such agents inside each cell, and that acting in concert they might account for the impressive feats of chemistry which living organisms accomplish. The few who took such an idea seriously during the nineteenth century did so more as a matter of prophetic vision than of scientific fact.

It was largely the work on fermenting yeast juice which put enzymes squarely in the centre of the biochemical map—or rather, dotted them in strategic positions all over it. Out of this work has developed a phase of intensive research on enzymes. In the sixty

LIFE'S BASIC DEVICE—SPECIFIC CATALYSIS BY ENZYMES

years following the Buchners' discovery, the number of known enzymes grew from a mere handful to pass the 600 mark—and the rate at which new enzymes are being discovered still shows no sign of slackening.

The word "enzyme" literally means "in yeast". Daily on the lips of thousands of biochemists, it serves as a constant reminder to them of the debt their science owes to the study of alcoholic fermentation.

IV

Organization and Efficiency—Subcellular Particles and Biological Oxidation.

Care in handling again—preparing subcellular particles

PEERING through a microscope, the student of elementary biology can hardly fail to be impressed by the dark-staining cell nucleus. Here, it seems, the important business of life must go on, and the pale cytoplasm looks like just a flabby, uninteresting surround. The impression is only partly justified. True, the nucleus contains the genetic material which carries the master-plan from generation to generation (Chap IX); but more of the machinery which hums with the day-to-day activity of cells is found outside it.

The cytoplasm itself is not without microscopic structure. In it there are many mitochondria, with dimensions of the order of 1 μ (one-thousandth of a millimetre), not far above the limit of the resolving power of ordinary light microscopy ($\frac{1}{4}$ μ). (To fix a point on the microscopic scale of magnitude, it is useful to remember that human red blood corpuscles are 8 μ in diameter.) The shapes and sizes of mitochondria vary somewhat; perhaps a typical case may be taken to be a rod, 2 μ long and $\frac{1}{2}$ μ wide.

Electron microscopy reveals still smaller structures. A system of membranes runs through the cytoplasm, and to it are attached particles called ribosomes, with diameters of 100 to 150 Å (Ångström units). Since 10,000 Å make 1 μ, the ribosomes are two orders of magnitude smaller than mitochondria. Their dimensions approach those of very large molecules; bond lengths between covalently linked atoms of carbon, oxygen and nitrogen fall within the range 1·2 to 1·6 Å, so that the diameter of a ribosome is about a hundred such bond lengths. Within the mitochondria, too, the electron microscope shows finer structure. Each is surrounded by a double membrane, and running most of the way across its width is a further series of double membranes which are probably formed by the inner member of the boundary membrane being thrown into high folds or "crests" (Fig. 12).

Methods have been developed for getting preparations of subcellular particles in quantities large enough to find out what

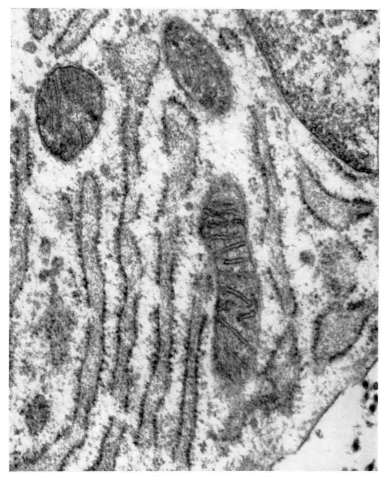

Fig. 12. Electron microscope photograph of a portion of a guinea-pig fibroblast (a connective tissue cell).

The cell membrane runs across the right lower corner, and at top right is a portion of the nucleus. In the cytoplasm can be seen three mitochondria with the characteristic paired membranes of their "crests". The ribosomes appear as black dots on the intra-cytoplasmic membranes. These membranes form flattened vesicles, the contents of which appear in this case denser than the cytoplasmic matrix; it is uncertain whether the breaks that can be seen in the membranes are functionally significant or are artifacts produced during the preparation of the sections. Ribosomes are abundant particularly in cells that synthesize much protein; fibroblasts form fibres of the protein collagen, the most characteristic constituent of connective tissues.

Magnification, 60,000 ×; stained with phosphotungstic acid. (Photograph supplied by Dr. J. A. Chapman).

ORGANIZATION AND EFFICIENCY

biochemical tricks they can perform in the test tube. The cells are broken in a "homogenizer", a special kind of test tube with a closely fitting glass or plastic plunger. Tissue fragments caught in the narrow space between the wall of the tube and the moving plunger are subjected to a shearing stress strong enough to break most of the cells. Many of the nuclei and other particles remain intact and can now be separated according to their sizes and densities by differential centrifuging—that is, by spinning slowly at first to make the heavier particles settle, then faster to sediment the lighter ones. The fractions so obtained consist of different types of cell fragments. Correlating them with the cell structures shown by various forms of microscopy naturally presents problems; but conditions for homogenizing and centrifuging have been worked out for which such correlations can meaningfully be made.

So much of the most significant biochemical work is done nowadays with subcellular particles that it is worth discussing their preparation in some detail. The underlying considerations raise some points significant for biochemical method in general.

All through the procedure it is, of course, important to keep conditions as mild as possible, so that damage to the particles is minimized. If the particles are damaged they may, instead of carrying on business as usual, perform biochemical tricks quite different from those they perform in the intact cell. These particles are even larger and more labile than protein molecules in general are, and in addition they are still metabolically active, so that even more care has to be taken in handling them, Such care is, of course, crucial for much of what biochemists do. If explanation in terms of smaller parts (p. 16) is to be valid, the parts must be studied in a state as near the native as is feasible. The aim must be to get them out with such dexterity and expedition that they hardly realize, so to speak, that they are no longer snugly at home inside their accustomed cells.

The story goes that the Arabs could cut a sleeping man's head off so neatly that, when he awoke, he was not aware that anything had happened. The ideal in preparing subcellular particles is very similar—and, of course, equally impossible to attain. Critics have objected that particles cannot be the same when isolated as they were *in vivo*, in the midst of the complex network of metabolic interrelationships. From the practical point of view, however, the study of those relationships is hardly possible without experiments on the properties and interactions of particles isolated by the mildest possible means, exercising continuous critical judgment to decide which effects are pure artifacts and which have some bearing on the biological reality. The objection becomes pedantic and obstructionist when pushed to extremes, but it does serve as a valuable caution against glib interpretation and as a reminder of the need for the utmost gentleness in preparation. (The general point is taken up again on p. 92.)

It goes without saying, then, that exposure of subcellular particles to any but the mildest reagents must be avoided. Even in media of the most favourable composition, however, many of their metabolic activities are lost or changed within an hour or so at 37°C. Getting information from them has been compared with straining to catch the last words falling from the lips of a dying man. Since it is not in general feasible to carry the preparation through in less than a few hours, degenerative changes are slowed down by cooling to near the freezing point. The particles are put into a state of artificial hibernation, as it were; only when the separation is completed is animation revived by warming to 37° in order to study the chemical reactions they are capable of performing.

Cooling, indeed, is the biochemist's most general answer to the general problem of changes during handling. Whatever the nature of the changes that occur, the lower the temperature the slower they are. If the typical procedure of chemistry is to mix and heat, that of biochemistry is to cool and separate. Means of cooling are, accordingly, as prominent a feature of biochemical laboratories as means of heating are in the usual image of a chemical laboratory. Normal provisions include not only liberal supplies of ice, but also refrigerated centrifuges and cold rooms kept at about $+2°C$, equipped with such things as benches and electricity supply so as to offer working as well as storage space. Facilities for long-term storage at about $-20°C$ (as for frozen food) are also usual, but the processes of freezing and thawing themselves may damage particles and large molecules and often have to be avoided.

Rat liver is a favourite tissue from which to prepare subcellular particles. Immediately after the rat has been killed—by cutting off the head, for instance—the liver is dissected out and cooled in ice; all the subsequent operations are performed only just above the freezing point. Pieces of liver are homogenized in a suitable medium, often an aqueous salt or sucrose solution. A typical centrifuging procedure might be as follows. First, the homogenate is spun for ten minutes at 700 g (i.e., 700 times the force of gravity) to bring down nuclei and any intact cells or large particles that have survived homogenization. The supernatant is then further spun for ten minutes at 5,000 g to sediment the mitochondria, and from the new supernatant, finally, a fraction of particles called microsomes are brought down in one hour at 50,000 g. Such a centrifugal force cannot be attained in a simple centrifuge; the instrument used is a kind of ultracentrifuge, in which the space around the rotor is evacuated to lessen air resistance. (It need not be quite as complicated as the kind of ultracentrifuge used for determining protein sedimentation rates (p. 50), since it does not need an optical system for recording boundary positions; a preparative—as distinct from an analytical—ultracentrifuge is used here. It may be of interest to point out that centrifugal force is proportional to ωr^2, where ω is the angular velocity and r the radius; to achieve 50,000 g, a rotor for which r is 8 cm must spin at nearly 24,000 revolutions per minute.)

There is no guarantee, of course, that any one fraction obtained by the centrifuging procedure contains only one type of particle. In the mitochondrial fraction, for instance, have been detected particles other than the mitochondria themselves. They have been called lysosomes; typically rather smaller, they can be partially separated from the mitochondria by further, more refined differential centrifuging.

The microsome fraction consists of fragments of the cytoplasmic membranes still studded with ribosomes. Besides containing protein, these membranes (like those of the mitochondria) are relatively rich in fat-like components and can be solubilized by substances having detergent action, such as sodium deoxycholate. When this is done, the ribosomes are liberated and can be centrifuged down (in an hour at 100,000 g, for instance).

Cells as more than bags of enzymes—enzymes in the particles

In enzyme make-up, the various particle fractions differ from each other in quite characteristic ways. Nuclei, while not destitute of enzymes, are in general rather poor in them. Mitochondria, on the other hand, contain highly active oxidizing systems, and are responsible for much of the oxidative activity of cells. Lysosomes carry a number of hydrolytic enzymes capable of attacking and destroying cell constituents. Ribosomes are centres for the important process of protein synthesis (p. 176). The cell sap, the non-particulate fraction which remains as supernatant even after the hardest centrifuging, contains in free solution the enzymes responsible for the anaerobic breakdown of carbohydrate.

Cells are thus biochemically organized in a way rather more complex than was at one time thought. In that exhilarating flush of enthusiasm between the two World Wars, when the route of sugar breakdown to alcohol or lactate was being elucidated, optimists felt that an understanding of intracellular metabolism required only the discovery of enough enzymes to catalyse each of the many steps in long reaction sequences. No organization seemed necessary beyond that inherent in enzyme specificity—the specificity of one enzyme being such as to act on the product of another enzyme, turning it into a substance susceptible to a third enzyme, and so on until the end product is obtained. Cells, however, are not just bags of enzymes. Only some of their enzymes float around randomly in free solution. The rest are organized into blocks, rather as a factory might be, in the interests of safety and efficiency.

Lysosomes are packets of destructive enzymes which it seems important to keep enclosed and separate from the other cell constituents during active life. Probably it is only with the death of the cell, when autolysis ensues, that these enzymes come into full operation to mop up the debris.

In the mitochondrial membranes, it seems, some of the enzymes catalysing oxidative reaction sequences are arranged in spatial sequences corresponding to the functional ones, on the production line principle. If this is so, each unit of the catalytic machinery, instead of having to diffuse randomly until by chance it meets its appropriate reaction partners, is fixed in place between those preceding and succeeding it in order of action. Certainly, some of the catalysts concerned with oxidation are firmly attached to the mitochondrial membranes—so firmly that it is difficult to find means of prizing them off without destroying them. This fact on the one hand represents a major experimental difficulty in studying them; on the other hand, it may well be highly significant in itself. The catalysts may be organized into functional assemblies "built in, or woven into, the texture of the mitochondrial membranes in the same manner as repeated decorative patterns are woven into a sheet of damask", as Palade, a leading electron microscopist, has put it. Many mitochondrial catalysts are not firmly attached to the membranes, but seem to be present in solution inside the mitochondria. Even in this case, however, the efficiency of a set of collaborating catalysts is presumably improved by concentrating them within the compact mitochondria instead of letting them wander freely through the much larger space of the whole cytoplasm.

It is by the oxidative activities of the mitochondria that the breakdown of carbohydrate, begun anaerobically (Chap III) by the enzymes of the cell sap, is completed to yield carbon dioxide and water. The reactions involved are considered in more detail below, taking them in two groups. One group is that by which the carbon skeleton is broken down. Since the end-product is carbon dioxide, this process must of course be oxidative, but molecular oxygen does not participate directly; instead, pairs of hydrogen atoms, removed in several of the steps which lead to complete breakdown, are handed to hydrogen acceptors. From one acceptor they are passed to another in a series which eventually—but only after some four or more intermediates—ends with oxygen. Such a series is called a "respiratory chain", and its operation constitutes the other group of reactions to be considered.

Passing the buck—respiratory chains

One puzzle immediately springs to mind in connection with respiratory chains. Why do in five steps what could be done in one? Why pass the buck instead of getting on with the job? Extra steps require extra catalytic machinery; and natural selection, one feels confident, should by now have pruned any gross extravagances in basic biochemical designs. A dozen steps between glucose and alcohol are no extravagance, because the chemical change is so far-reaching that it could hardly be accomplished in much less. To combine hydrogen atoms with oxygen to form water, however, is something that could be done directly, and the insertion of four or more intermediate steps seems purely gratuitous at first sight.

Later, it will be possible to offer fuller justification for the multistage mechanism (p. 152). For the moment, it is enough to bear in mind that oxidation is not, after all, an end in itself; to get rid of food and oxygen, and to form carbon dioxide and water, do not in themselves contribute to the maintenance of life. Cells oxidize to get a supply of useful energy, and performing oxidations stepwise makes it possible to trap a high proportion of the energy released in a form available for use in biological processes—that is, as ATP.

As regards the precise nature of all the carriers constituting the chains, knowledge is as yet far from complete. (Some of the experimental difficulties that stand in the way of their elucidation are mentioned on p. 94 and p. 103.) Knowledge remains vague especially as regards the middle portion of the chains; beginning and end are relatively well understood.

The usual beginning is NAD, which has already been mentioned for its role in the anaerobic phase of carbohydrate breakdown. In some cases, NAD is replaced by its phosphate, called NADP. (Old names for NADP are triphosphopyridine nucleotide or TPN and coenzyme II or CoII; these correspond to the old names for NAD mentioned on p. 65.) In a few other cases, compounds of an entirely different type, derivatives of riboflavin, are involved.

At the end of the chains, as oxygen is approached, are carriers called cytochromes. These, as their name suggests, are intracellular pigments. Chemically, they are conjugated proteins quite closely related to the blood pigment haemoglobin—that is, they consist of proteins attached to which are non-protein moieties or prosthetic groups of haem. A haem consists of a rather complex, nitrogen-containing, heterocyclic structure called a porphyrin which holds in its centre an atom of iron. Various haems occur naturally, with different substituents in the porphyrin portion. Aerobic cells of all types contain a number of cytochromes, differing from each other in the exact structures of the porphyrin as well as the protein components. All cytochromes, however, contain iron and function as carriers in basically the same way—namely, by alternating between the oxidized or ferric and the reduced or ferrous forms. Writing Cyt for the protein and porphyrin components, and e for an electron,

$$Cyt \cdot Fe^{+++} + e \rightleftharpoons Cyt \cdot Fe^{++}$$

It is necessary to remember, of course, that from the oxido-reduction standpoint removal of two hydrogen atoms, removal of two electrons and addition of one oxygen atom are all equivalent. Thus the above reaction could equally well be written

$$Cyt \cdot Fe^{+++} + [H] \rightleftharpoons Cyt \cdot Fe^{++} + H^+$$

Oxidation of a cytochrome by oxygen takes place as follows:—

$$4Cyt \cdot Fe^{++} + 4H^+ + O_2 \rightarrow 4Cyt \cdot Fe^{+++} + H_2O$$

(It is worth noting, incidentally, that haemoglobin acts in a different way. Here, the iron normally stays in the ferrous form and binds a molecule of oxygen as such in the form of a complex.

$$Hb \cdot Fe^{++} + O_2 \rightleftharpoons Hb \cdot Fe^{++} \cdot O_2$$

The theory behind committing suicide with a gas oven or a car exhaust is that carbon monoxide, CO, powerfully competes with O_2 in this reaction. Carbon monoxide is no oxidizing agent; in replacing oxygen here, it is clearly acting not by oxidizing but by forming a complex.)

The major respiratory chain of mitochondria can be represented in outline as follows:—

$$\begin{array}{l} AH_2 \searrow \\ BH_2 \rightarrow NAD \rightarrow X \rightarrow Cyt\ c \rightarrow Cyt\ a \rightarrow O_2 \\ CH_2 \nearrow \end{array}$$

The arrows here show the pathway of hydrogen or electron transport, *not* the conversion of one indicated substance into another. There is no question of turning AH_2 into NAD; the reaction implied is

$$AH_2 + NAD = A + NADH_2$$

AH_2 represents an intermediate in the breakdown of a foodstuff by a route that involves, as one of its constituent steps, the oxidation (i.e., dehydrogenation) of AH_2 to A. Actually, there are a number of such intermediates—BH_2, CH_2 etc.; in each case, an appropriate specific enzyme catalyses the transfer of a pair of hydrogen atoms to NAD. Such enzymes are collectively called dehydrogenases, and particular ones are distinguished by the names of the substrates they dehydrogenate; for instance, the enzyme which dehydrogenates lactate is called lactate dehydrogenase.

Exactly how the reoxidation of $NADH_2$ is effected remains unclear—that is, the chemical nature of the next carrier has not been decided, and it is far from certain even whether one or more carriers participate in this region of the chain (cf. p. 95). Skating over this controversy by glibly writing X, we have

$$NADH_2 + X = NAD + XH_2$$

and thus arrive at the cytochrome region.

A high proportion of cellular oxidation proceeds through at least two cytochromes, which are called *c* and *a* respectively. (The letters designating the different cytochromes are given according to the positions of their absorption bands, not the order in which they act. Cytochrome *b*, which occurs in virtually all aerobic cells with *a* and *c*, may act before *c* in the chain.) The last cytochrome in the chain (either cytochrome *a* or one like it) is distinguished by its ability to react directly with oxygen. Quite possibly, cytochromes other than the two shown are also involved, but confining attention to these two, the reactions would be:—

$2XH_2 + 4Cyt\ c \cdot Fe^{+++} = 2X + 4Cty\ c \cdot Fe^{++} + 4H^+$
$4Cyt\ c \cdot Fe^{++} + 4Cyt\ a \cdot Fe^{+++} = 4Cyt\ c \cdot Fe^{+++} + 4Cyt\ a \cdot Fe^{++}$
$4Cyt\ a \cdot Fe^{++} + 4H^+ + O_2 = 4Cyt\ a \cdot Fe^{+++} + 2H_2O$

It will be noticed that all the H^+ formed in the first of this group of three reactions is used up again in the last. There is, therefore, no ground for the sudden fear that strikes some students that the continual reduction of a cytochrome by XH_2 will send their intracellular pH plummeting downwards to lethal ranges.

Into the neck of the funnel—formation of acetylcoenzyme A

The reactions by which carbon dioxide is formed by complete oxidation of carbohydrate follow on from the anaerobic reaction sequence described earlier (Chap. III). As the true starting point for the aerobic phase it is proper to take pyruvate, since its reduction to lactate (reaction 13 of Fig. 11) does not take place when the oxygen supply is plentiful. It will be remembered that the significance of this reaction lies in the fact that from $NADH_2$ it regenerates the NAD necessary for the continued oxidation of glyceraldehyde phosphate (reaction 6). Under fully aerobic conditions, however, $NADH_2$ is reoxidized via a respiratory chain and is therefore not available for reducing pyruvate. Quite on the contrary, any lactate that might have accumulated during a period of oxygen lack can be reoxidized to pyruvate by reversal of reaction 13. (This reaction is catalysed in both directions by the same enzyme, lactate dehydrogenase; enzyme-catalysed reactions in general are freely reversible except where energy considerations (p. 146) forbid).

The route that leads to complete oxidation of pyruvate begins with a decarboxylation, but not a straightforward one like that which yields acetaldehyde during alcoholic fermentation by yeast. Different enzymes can awaken different latent potentialities for reaction in the same substrate. The enzyme system which acts here catalyses a decarboxylation complicated by two factors. First, it involves also an oxidation at the expense of NAD to the level of acetic acid. Second, the acetic acid does not arise in the free state but in combination with a substance called coenzyme A. This substance contains an —SH or sulphydryl group with which the acetic acid is condensed, giving acetylcoenzyme A. Writing CoA·SH for coenzyme A, the overall result of the oxidative decarboxylation of pyruvate is

$$CH_3COCOO^- + H^+ + NAD + CoA \cdot SH = CoA \cdot S \cdot COCH_3 + NADH_2 + CO_2$$

Reoxidation of $NADH_2$ is effected, of course, via a respiratory chain.

One-third of the carbon atoms of carbohydrate have now been turned into carbon dioxide; the remaining two-thirds are in the form of the acetyl groups of acetylcoenzyme A. These acetyl groups are themselves finally oxidized to carbon dioxide by a reaction sequence known as the citrate cycle. They occupy a key position in catabolism as a whole, for the breakdown of carbohydrate via pyruvate is not the only route by which they can arise. The degradation of the long straight chains of fatty acids, which are the major constituents of natural fats, takes place by successively knocking off pairs of carbon atoms; the two-carbon fragments which result are the acetyl groups of acetylcoenzyme A. Many of the carbon atoms of the amino-acids of proteins, too, eventually turn up as these acetyl groups.

The catabolism of all the main classes of foodstuffs thus converges to a large extent on acetyl in the form of its compound with coenzyme A. Further oxidation of these two carbon atoms to carbon dioxide can take place by reactions (those of the citrate cycle) which are common to the breakdown of carbohydrates, fats and proteins. The catabolic flow-diagram is funnel-like—wide at the beginning, to catch the considerable chemical diversity of foodstuffs, narrow at the end, where the routes join in a final common pathway. Acetylcoenzyme A stands at the entrance to the neck of the funnel. (Fig. 13). The funnelling device amounts to an approach to, metabolic standardization; it makes possible an appreciable saving in the diversity of catalytic machinery with which cells have to equip themselves. From the standpoint of economy and efficiency, this is a strong point to set against the insinuation of extravagance that was made above in connection with respiratory chains.

Two different but related efficiency-increasing devices are thus recognizable in the way the oxidative breakdown of foodstuffs is engineered. One concerns the nature of the steps involved, the other the spatial arrangement of the machinery responsible for bringing them about. By making the breakdown routes converge, the range of catalytic machinery required is reduced; this limited range is concentrated in the mitochondria, facilitating co-operation by vicinity. Several reactions of the citrate cycle involve dehydrogenation, and respiratory chains are at hand in the mitochondria to deal with the hydrogen removed. Some of the carriers making up the later stages of the respiratory chains are firmly embedded in the mitochondrial membranes, suggesting that they may be actually fixed in sequence on the production line principle. (Moreover, the respiratory chains themselves are the necks of metabolic funnels. Here, the gaping mouths of the funnels are formed by the various dehydrogenases collecting hydrogen from their respective substrates; these substrates are represented by AH_2, BH_2 etc. on p. 78 and include citrate cycle intermediates.)

The last round of catabolism—the citrate cycle

The citrate cycle is the last round of catabolism; it finally ejects

THE BIOCHEMICAL APPROACH TO LIFE

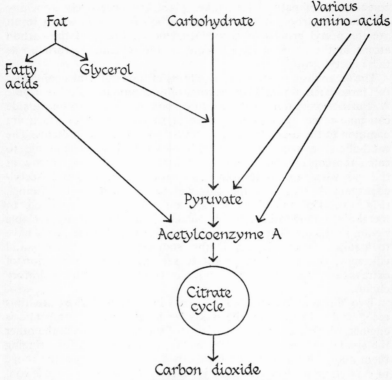

Fig. 13. *Catabolic convergence*

The major breakdown routes converge funnel-like on the acetyl groups of acetylcoenzyme A, which are then oxidized to carbon dioxide by the citrate cycle.

Fatty acid chains are chopped into two-carbon units which form the acetyl groups of acetylcoenzyme A. Glycerol is convertible (by phosphorylation and dehydrogenation—cf. p. 66) into dihydroxyacetone phosphate, which lies on the route from carbohydrate to pyruvate (Fig. 11). Much of the carbon of the amino-acids of proteins is also convertible into acetyl, either via pyruvate or otherwise.

the carbon atoms as carbon dioxide. Clearly, its operation is of the utmost importance, since it is capable of completing the oxidation of carbohydrates, fats and most of the carbon of amino-acids. It was Krebs, working at that time in Sheffield, who in 1937 put forward the idea of such a cycle, using a number of reactions already known and suggesting a few more to complete the merry-go-round.

ORGANIZATION AND EFFICIENCY

Krebs shared the Nobel prize for medicine in 1954 and the cycle is still often known by his name. The enzymes that catalyse each of the reactions that make it up have been purified and studied in considerable detail.

Basically, what happens is this. The acetyl group of acetylcoenzyme A condenses with oxaloacetate to give citrate, which is then degraded by a series of reactions to regenerate oxaloacetate. This can then condense with more acetylcoenzyme A, so that the cycle provides a mechanism for the breakdown of unlimited quantities of acetyl groups. The coenzyme A liberated during citrate formation is available for combining with fresh acetyl groups and carrying them into the cycle.

Two of the reactions leading from citrate to oxaloacetate involve elimination of carbon dioxide, accounting for both the carbon atoms of the acetyl group. Four reactions involve removal of a pair of hydrogen atoms each; these are turned into water through the operation of respiratory chains, ultimately involving four atoms of oxygen. The requirements for the complete oxidation of acetic acid (of which acetylcoenzyme A is a bound form) are thus satisfied, for the overall equation is

$$CH_3COOH + 2O_2 = 2CO_2 + 2H_2O$$

The reactions of the cycle are shown in full in Fig. 14. It is worth emphasizing again that, with these reactions, the complete oxidation of glucose has now been accounted for. The sugar is broken down to pyruvate as shown in Fig. 11; pyruvate is oxidatively decarboxylated to give acetylcoenzyme A, the acetyl groups of which are finally converted into carbon dioxide via the cycle.

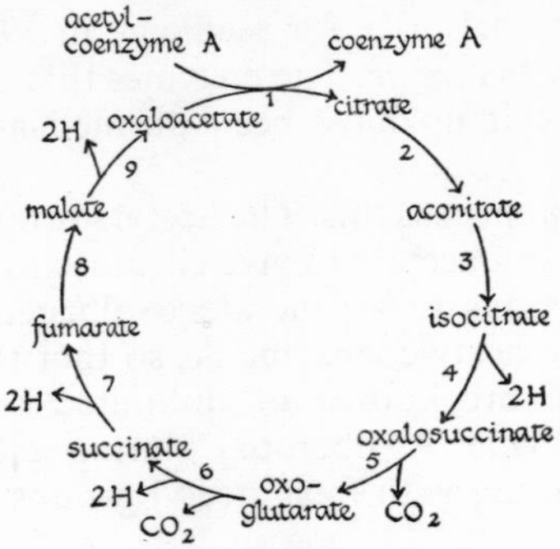

THE BIOCHEMICAL APPROACH TO LIFE

Fig. 14. *The citrate cycle*

All the intermediates of the cycle are the ions of di- or tricarboxylic acids. Oxaloacetate belongs to the former category, citrate to the latter; the conversion of the one into the other (step 1) in effect consists of the addition of an acetic acid moiety, acetylcoenzyme A being a reactive form of acetic acid. (Oxaloacetic acid itself derives its name from its formation, in theory, by condensing acetic acid with oxalic acid, $(COOH)_2$.)

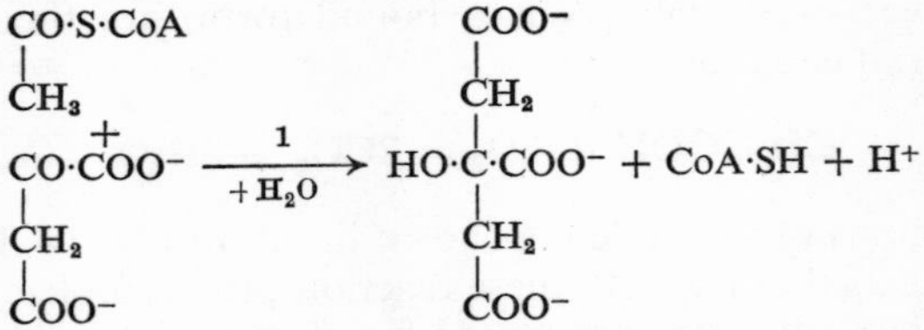

$$\begin{array}{c} CO \cdot S \cdot CoA \\ | \\ CH_3 \\ + \\ CO \cdot COO^- \\ | \\ CH_2 \\ | \\ COO^- \end{array} \xrightarrow[+H_2O]{1} \begin{array}{c} COO^- \\ | \\ CH_2 \\ | \\ HO \cdot C \cdot COO^- \\ | \\ CH_2 \\ | \\ COO^- \end{array} + CoA \cdot SH + H^+$$

The interconversion of citrate, aconitate and isocitrate (steps 2 and 3) is catalysed by a single enzyme, aconitase. Aconitate is an unsaturated compound; the enzyme, which catalyses the addition of the elements of water across the double bond, is indifferent which side gets the H and which the OH. Like the majority of metabolic reactions, this one is freely reversible, and if the system were left to itself, an equilibrium mixture of the ions of the three tricarboxylic acids would result. In practice, however, the continuous removal of isocitrate pulls the reactions over to the right.

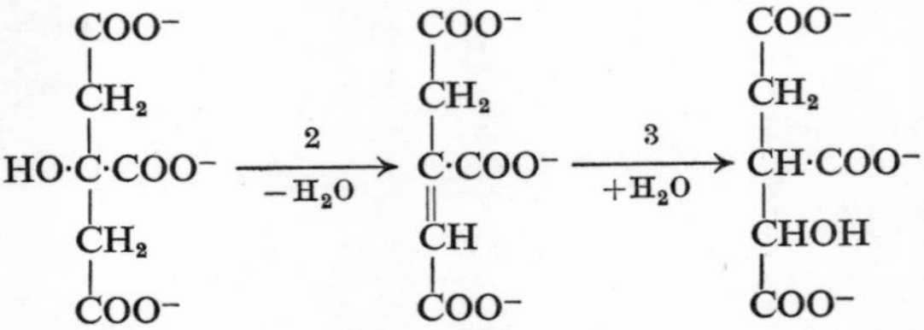

$$\begin{array}{c} COO^- \\ | \\ CH_2 \\ | \\ HO \cdot C \cdot COO^- \\ | \\ CH_2 \\ | \\ COO^- \end{array} \xrightarrow[-H_2O]{2} \begin{array}{c} COO^- \\ | \\ CH_2 \\ | \\ C \cdot COO^- \\ || \\ CH \\ | \\ COO^- \end{array} \xrightarrow[+H_2O]{3} \begin{array}{c} COO^- \\ | \\ CH_2 \\ | \\ CH \cdot COO^- \\ | \\ CHOH \\ | \\ COO^- \end{array}$$

Isocitrate is removed (step 4) by a dehydrogenation at the newly formed hydroxyl group. The hydrogen acceptor is NADP and the enzyme which catalyses the reaction is called isocitrate dehydrogenase.

$$\begin{array}{c}\text{COO}^-\\|\\\text{CH}_2\\|\\\text{CH·COO}^-\\|\\\text{CHOH}\\|\\\text{COO}^-\end{array} + \text{NADP} \xrightarrow{4} \begin{array}{c}\text{COO}^-\\|\\\text{CH}_2\\|\\\text{CH·COO}^-\\|\\\text{CO}\\|\\\text{COO}^-\end{array} + \text{NADPH}_2$$

Straightforward decarboxylation of oxalosuccinate (step 5) is catalysed by oxalosuccinate decarboxylase, leaving α-oxoglutarate (also called α-ketoglutarate); this substance is named as a derivative of glutaric acid, COOH·$(CH_2)_3$·COOH, the next higher homologue of succinic acid in the dicarboxylic acid series.

$$\begin{array}{c}\text{COO}^-\\|\\\text{CH}_2\\|\\\text{CH·COO}^-\\|\\\text{CO}\\|\\\text{COO}^-\end{array} + \text{H}^+ \xrightarrow{5} \begin{array}{c}\text{COO}^-\\|\\\text{CH}_2\\|\\\text{CH}_2\\|\\\text{CO}\\|\\\text{COO}^-\end{array} + \text{CO}_2$$

Oxoglutarate is decarboxylated in its turn (step 6) but, unlike the preceding step, this is not a simple but an oxidative decarboxylation. It involves the reduction of NAD, just like the similar reaction involving pyruvate (p. 80)—which, in fact, it closely resembles in mechanism. The structure undergoing reaction in both cases is that of an α-oxoacid, RCH_2·$COCOO^-$, where R is —CH_2COO^- for oxoglutarate and H for pyruvate. Just as, in the case of pyruvate, the product is acetylcoenzyme A rather than free acetate, so in the case of oxoglutarate it is succinylcoenzyme A. The coenzyme A is subsequently split off, however, so that the overall result is to leave succinate.

$$\begin{array}{c}\text{COO}^-\\|\\\text{CH}_2\\|\\\text{CH}_2\\|\\\text{CO}\\|\\\text{COO}^-\end{array} + \text{H}_2\text{O} + \text{NAD} \xrightarrow{6} \begin{array}{c}\text{COO}^-\\|\\\text{CH}_2\\|\\\text{CH}_2\\|\\\text{COO}^-\end{array} + \text{NADH}_2 + \text{CO}_2$$

Succinate is dehydrogenated (step 7), creating a double bond. The enzyme is succinate dehydrogenase and the hydrogen acceptor, in this case, is neither NAD nor NADP but a riboflavin derivative, denoted here as F.

$$\begin{array}{c}\text{COO}^-\\|\\\text{CH}_2\\|\\\text{CH}_2\\|\\\text{COO}^-\end{array} + \text{F} \xrightarrow{7} \begin{array}{c}\text{COO}^-\\|\\\text{CH}\\||\\\text{CH}\\|\\\text{COO}^-\end{array} + \text{FH}_2$$

THE BIOCHEMICAL APPROACH TO LIFE

The two reactions by which oxaloacetate is formed from fumarate are analogous to the two by which oxalosuccinate is formed from aconitate; as with the analogy between pyruvate and oxoglutarate, the difference is a carboxymethyl group, for aconitate can be regarded as fumarate substituted by —$CH_2.COO^-$. In the first reaction, the double bond is hydrated (step 8, resembling 3), then the resulting alcoholic group is dehydrogenated (step 9, resembling 4). The enzymes are called fumarase and malate dehydrogenase respectively.

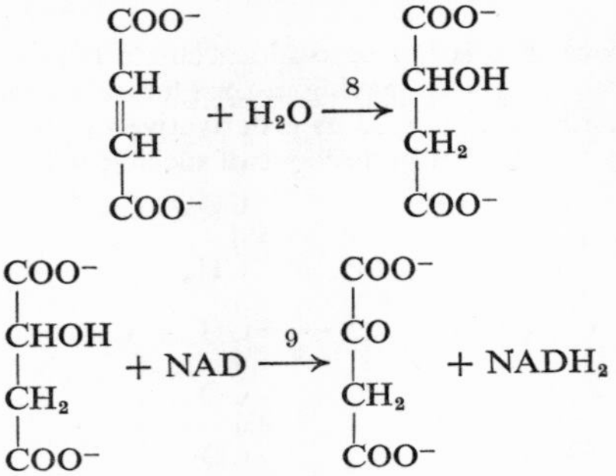

$$\begin{array}{c} COO^- \\ | \\ CH \\ || \\ CH \\ | \\ COO^- \end{array} + H_2O \xrightarrow{8} \begin{array}{c} COO^- \\ | \\ CHOH \\ | \\ CH_2 \\ | \\ COO^- \end{array}$$

$$\begin{array}{c} COO^- \\ | \\ CHOH \\ | \\ CH_2 \\ | \\ COO^- \end{array} + NAD \xrightarrow{9} \begin{array}{c} COO^- \\ | \\ CO \\ | \\ CH_2 \\ | \\ COO^- \end{array} + NADH_2$$

Notes for Fig. 14

The various NAD molecules reduced are reoxidized, of course, by the respiratory chain mechanism. Reoxidation of NADP probably occurs by very similar means, and the riboflavin derivative reduced in step 7 is dealt with by a somewhat modified respiratory chain.

V

Two Approaches to Biological Explanation— Analogy and Analysis

The biochemist's dilemma—to imitate or to take to pieces?

THE stage is now set with enough facts for a rather more searching examination than could be attempted earlier (Chap. I) of the kind of explanation of biological phenomena that biochemistry can (or should) attempt to offer.

In this chapter are set out two different views of how chemistry can be applied to biology. These views depend on different general principles which may be called *analogy* and *analysis* (these names having at least the virtue of intense alliteration). By analogy is here meant the comparison of living systems to models made from non-biological constituents. Analysis is taken in its most literal sense, to mean dividing things into spatially smaller pieces. (This approach is often called "atomistic", but the term seems inappropriate here because of its association with one particular type of smaller piece, the atoms of chemistry.)

The choice offered by these two alternatives is whether to make models to imitate living organisms, or to start with real living organisms and take them to pieces. To put it crudely, should the biochemist fill his test tubes with reagents from his shelf in attempts to simulate the phenomena of life, or with the mashed-up debris of living things? The issue is a big one, and its implications extend to experimental biology in general.

The limitations of analogy. Descartes' animal machine and its successors

Model-making is recognized as an important general procedure in science. Of course, the models need not always be physically constructed with sticks, string, sealing-wax and the like; they may be theoretical models—idealizations of a situation for purposes of calculation, for instance. The particular kind of model that first springs to mind in connection with biochemistry is the non-living model used as an analogy to a living system, and the discussion here is restricted to this particular type of model and this particular type of analogy.

Since the rise of mental coherence, men have used analogies to try to bridge the gulf between living and non-living; but they have not always drawn the analogies in the direction which nowadays seems obvious to most people—arguing from non-living to living, that is. The principle behind the use of analogies as explanations is the old one that, to advance knowledge, one must proceed from the known to the unknown. To give satisfaction as an explanation, a model must seem easier to understand than the situation it sets out to clarify. In antiquity and in the Middle Ages, life seemed no more mysterious than inorganic nature—if anything less so, being closer to the thinker's immediate experience. Accordingly, it seemed at least as appropriate to explain the non-living in terms of the living as the other way round. The fall of heavy bodies was explained as due to their tendency to return to their "natural" positions below, as though they were home-sick animals; and metals were thought of as "growing" under the earth.

To the twentieth century, such explanations seem crude and essentially anthropomorphic. Are analogies in the reverse direction more acceptable?

It was in the seventeenth century that conditions became such as to make it obvious to argue from non-living to living. The decisive change was the development of mechanics—notably the formulation of laws of dynamics, such as Galileo's mathematical treatment of the motion of falling bodies. Inanimate matter and its motion now seemed wonderfully comprehensible—and the incentive to explain the phenomena of life in terms of them became correspondingly irresistible.

The most influential attempt to do so was made by the French philosopher Descartes. His book *On Man* (published in 1662 but written thirty years earlier) has been called the first book dedicated to physiology as such; but a very peculiar physiology it was that it contained. It described a theoretical model of a man, constructed on the principles used by contemporary machines—"clocks, artificial fountains, mills and other machines which, though made by man, yet have the power of moving in various ways." Muscles were supposed to work by being inflated with a fluid from the brain via the nerves, making them shorter and wider. Nerves were pictured as containing tiny threads running from sense-organs to brain. On being moved by external stimuli, the threads were supposed to operate appropriate valves so that the nervous fluid was directed

TWO APPROACHES TO BIOLOGICAL EXPLANATION

into the right muscles for suitable movements to occur. By means of a sufficiently elaborate series of interconnections in the brain, the machine could be "programmed" for an elaborate series of reflex actions.

Descartes proposed that a mechanism of this type could account for all the movements of animals. (He did allow a proviso in the case of man, but not of animals, for the power of mind to direct those movements—a minority—under direct voluntary control.) The animal machines were compared to the water-actuated statues in the grottos of the royal gardens, which could be set into various kinds of motion by appropriate arrangements of the tubes leading water to them. "External objects acting on the sense-organs, and thereby leading to diverse movements, according to the disposition of the parts in the brain, are like strangers entering one of these grottos and thereby setting the statues into unthinking motion. For they cannot enter without treading on certain flag-stones so disposed that if, for instance, they approach a bathing Diana, they make her hide behind reeds; if they try to follow her, they are approached by a Neptune who menaces them with his trident; if they go some other direction, they bring out a marine monster which spews water in their faces; or similar things, according to the whim of the engineers who constructed the statues."

Descartes was not the first to think along these lines, but it was he who gave momentum to the craze of "mechanism" in biology—explaining the phenomena of life by effects observed in non-living systems. To him, as to later generations, this approach held the great attraction of a unified picture embracing both living and non-living nature. Unfortunately, Cartesian mechanist theory rode with rough-shod unconcern over the details of biological fact. Inevitably, it brought on itself the reaction of eighteenth century "vitalism", in the form of protests that living organisms must be looked at as they really are, not as they conceivably might be. Nineteenth century materialism, however, again found itself strongly attracted by the Cartesian approach to physiology. "Can ye make a model of it?" Lord Kelvin, the physicist, is supposed to have said, epitomizing a view common among Victorian scientists. "For if ye can, ye understand it, and if ye canna, ye dinna!"

Many different kinds of model have suggested themselves, and fashion in this respect has changed with the times. Where the seventeenth century thought in terms of simple mechanical devices, the nineteenth preferred heat engines. Nowadays, computers are in favour. These are essentially physical models; more closely relevant to the topic of this book are chemical ones, which take the form of *in vitro* reactions with overall results similar to *in vivo* ones, and hence in some measure imitating them.

Models are genuinely useful in so far as they represent theories about certain aspects of the functioning of living things, testable on material from living things. But it is not possible to get information about living organisms merely by studying artificial models. Simple and obvious though this fact may seem when so baldly stated, it can be obscured by the over-enthusiasm and paternal pride of the model-makers. If this happens, misdirection of effort may result.

The trouble with the Cartesian or any similar approach to physiology is not that the details of the models may not be right; even if they are wrong they may, like any theory, suggest useful experiments. The real fault lies in the basic assumption of the intellectual procedure—that a biological process is in some way "explained" when a model to imitate it has been thought of or constructed (cf., however, p. 108). Genuine explanation, with an increase in knowledge and understanding of life, is not achieved until the imitation has been critically compared with the biological reality. Quite recently, in mid-twentieth century, there has been talk of "synthesis under physiological conditions", meaning the artificial synthesis of natural products under conditions of temperature and pH not beyond the physiological range. But physiological conditions are not, after all, defined only by temperature and pH. Syntheses in living organisms may (and usually do) proceed via intermediates quite different from those used in such artificial syntheses. Studies of this type show what *could* happen in living organisms, not what *does* happen.

Apart from their value in teaching as expository devices, analogies to models have furthered science most, perhaps, by stimulating the study of the models themselves and of the non-biological materials used in their construction. The great structure of modern organic chemistry is itself a most striking monument to such a stimulus at work. The name "organic chemistry" is a verbal fossil which recalls the history of the subject. Originally, it meant the chemistry of organized, living beings. Not until mid-nineteenth century did it come explicitly to refer to carbon compounds rather than to natural products (p. 69). (Biochemistry is thus the ancestor of organic chemistry, rather than the other way round as is sometimes supposed.) Although the study of artificial carbon compounds soon acquired justifications of its own, in terms both of theoretical interest and of practical utility, it gained impetus originally from the desire to copy in the test tube the compounds and reactions of life.

TWO APPROACHES TO BIOLOGICAL EXPLANATION

Analogies applied to biological oxidations

Both the strength and the limitations of the approach to biochemistry via analogies are well illustrated in the study of biological oxidations (cf. also p. 104).

Of all the analogies that have ever been drawn between non-living and living, Lavoisier's comparison of burning and breathing was perhaps the most strikingly successful. "Respiration is a combustion," he wrote in 1783, "slow it is true but otherwise perfectly similar to that of charcoal." The similar dependence on air of the two processes had been shown a century before. Lavoisier, however, went much further by comparing them quantitatively, not only with respect to the carbon dioxide produced but also (in collaboration with the physicist Laplace) with respect to the heat evolved. What gives this work its importance is not the analogy itself, the idea of the similarity, but the critical comparison of the two processes, both subjected to direct experiment. But for this, Lavoisier could not be regarded as having laid the foundations for the study of metabolism, in terms both of material change and of energy transformation.

The obvious weakness of the comparison was the great disparity in the temperatures at which the two processes take place. Apart from invoking the general notion of catalysis (p. 56), scientists long remained at a loss how to account for this. In the later part of the nineteenth century, chemists became intrigued by the oxidizing powers of peroxides at ordinary temperatures, and studied them in model systems. This work added a good deal to our knowledge of peroxides, but not much to our knowledge of the major oxidation mechanisms of living cells.

A different kind of model turned out eventually to be more profitable. This kind was based on the catalytic effect of heavy metals on oxidations by molecular oxygen. Finely divided platinum and palladium act in this way, "activating" hydrogen atoms in organic substances to make them combine with oxygen. This effect was, during the second decade of this century, made the basis of a theory of biological oxidation which was widely accepted as the best available.

During the early nineteen-twenties, Warburg in Germany found better model systems. The catalysts here were certain charcoals, such as that obtained by charring blood. These contain a good many substances besides carbon; Warburg tracked their catalytic activity down to iron combined with something nitrogenous. They were better models in that they corresponded more closely to the biological systems in their behaviour and properties—notably in their rather similar susceptibilities to poisons such as cyanide in low concentrations. "Who could believe that this was only by chance in agreement with the behaviour of cell respiration?" enthused Warburg. His conclusion was that iron-containing catalysts operate during oxidation by cells. Fortunately, his intuition carried him where logic could not (possibilities other than iron were not ruled out); his guess was later confirmed by the discovery of cytochromes (p. 78).

Analogies to model systems, then, have played some part in bringing us our present knowledge of biological oxidations. Progressive refinements of the models, making them more and more faithful to the originals by more and more detailed checking and critical comparison, eventually led to a valuable hint about the participation of iron. It is worth remembering, however, that cytochromes were not actually discovered except by direct approach to living organisms. Their discovery was due to spectroscopic observations of a variety of cells (p. 100), followed by work on their isolation.

The objections to analysis

The view of biochemistry as the ultimate extension of biological analysis in terms of constituent parts has already been set out (Chap. I). In this aspect, it appears as the logical extrapolation of dissection to smaller magnitudes. Furthermore, it has been pointed out that form has a similar relevance to function at the molecular as at larger scales of magnitude; and biochemistry is concerned with the isolation of biological events, even more than with the isolation of biological materials (p. 53). In this aspect, it could perhaps be said to be the logical extrapolation of experiment by vivisection. Although it is not usually possible to keep an organism alive in a technical sense when constituents are isolated from it, care taken to minimize damage to molecules during isolation corresponds to the relative gentleness with which a whole animal is handled during vivisection; in each case, the aim is to maintain the entities under investigation functionally unaltered as far as possible.

All this sounds like an admirable programme for biochemistry. Explanation in terms of smaller parts is a kind of explanation which the human mind finds particularly satisfying, and it has proved time and time again to be particularly fruitful in practice. Unfortunately, this approach is not without its snags (p. 73). A time-honoured criticism by biologists of chemical operations, since the seventeenth century at least, has been that they are destructive. The implication of the word "organism" must not be forgotten; a living organism is organized, and its organization is a very important and characteristic thing about it. A method that has to destroy the organization before it can set to work cannot, therefore, claim to give a good explanation of life. To put it crudely, one cannot find out much about a rat by mashing it up, because there is then no rat left to find out about. The spatial and functional relationships between the parts are as important as the properties of the parts themselves.

The point is not merely that, by destroying the relationships, the chance is lost of finding out what they were. There is the further possibility that the behaviour of the parts themselves may now differ significantly from their natural behaviour. In the organism, the parts interact and influence each other in specific and delicate ways, and what they do depends on how they fit into the living machinery. Isolated, they may be made to do something different. Even supposing the parts not to have been substantially altered during isolation, the functional properties they can be made to display in the laboratory can be expected to include, besides those of physiological importance, others which are not called into play during life. How is one to tell which of the potentialities revealed by test tube events correspond to actual biological events? It may even be possible to assemble a number of test tube events into a series which fits together in an apparently convincing way but nevertheless bears little resemblance to any metabolic pathway operating inside cells. If one took a sledge-hammer to a washing-machine, it is conceivable that from the pieces one might construct a working toy train—but that should not be taken as evidence that a toy train was going round and round inside the intact washing-machine.

Circumspection is called for, then, before an effect observed *in vitro* is interpreted in terms of an *in vivo* process. To take one type of case that is very familiar in biochemistry—the fact that an enzyme catalyses a certain reaction in the test tube does not mean that the same reaction goes on in the living organism from which the enzyme came. Even if the enzyme has not been functionally altered during isolation, it may act on some other substrate *in vivo*. Catalysis by enzymes may be remarkably specific by comparison with other kinds of catalysis, but it is not absolutely specific; a given enzyme is usually capable of acting on a range of substrates united by some chemical resemblance—sometimes a narrow range defined by a close resemblance, sometimes a broader range. Any particular substrate tested *in vitro* may be an unnatural one which the enzyme does not encounter except through the machinations of biochemists. An enzyme, it might be said, engages in incidental hobbies when given the chance, as well as the significant work it does during normal life. To build up a picture of the functioning of intact organisms, the hobbies must be distinguished from the work.

Sometimes this is easy, when the substrate is an obviously artificial one, such as the synthetic dyes which some oxido-reduction enzymes can reduce. In other cases the answer is far from obvious. Given a number of substrates, among which it is not clear which is natural and which is not, the efficiencies with which the enzyme acts on them form no basis on which to judge; the enzyme may be more adept at its hobby than at its work—plenty of such cases are known. Even if a substrate can be shown to be present in the same tissue or cell as the enzyme, this is no guarantee that it will normally be acted on. An inhibitor may prevent the enzyme from working, or physical access of enzyme and substrate to each other may be impeded by the membranes of subcellular structures like mitochondria—for a cell is not just a bag in which enzymes and substrates mix indiscriminately (p. 75). Nor can it be assumed that a series of enzyme-catalysed reactions is physiological because they fit together neatly into some plausible functional pattern or metabolic pathway; that would be like taking the satisfactory working of the toy train as evidence that it pre-existed in the washing-machine.

Some of these difficulties are illustrated in the attempts that have been made since the nineteen-thirties to complete our knowledge of respiratory chains by identifying the carriers that operate between NAD or NADP and cytochromes (that is, to elucidate the nature of X in the respiratory chain written on p. 78). There have been hints that X may be of the nature of a flavoprotein. The flavoproteins are a group of conjugated proteins, the protein portions of which are attached to yellow prosthetic groups derived from riboflavin (Table IV). Several flavoproteins have been obtained which can oxidize $NADH_2$ by themselves accepting pairs of hydrogen atoms. Allocating them proper physiological roles, however, has been a thorny problem.

The difficulties concern the mechanism by which the flavoproteins are themselves reoxidized, enabling them to function catalytically as members of respiratory chains. The earliest "yellow enzymes" to be studied were obtained by Warburg (p. 91); they were oxidizable directly by molecular oxygen. There is, however, excellent evidence that a large proportion of cellular oxidation proceeds via cytochromes, and it is now generally assumed that these "yellow enzymes" were artifacts produced by damage during isolation to flavoproteins which originally were not autoxidizable. Later, flavoproteins were prepared which could not react directly with oxygen but could reduce certain synthetic dyes like methylene blue; they did not, however, react with cytochromes or any other known physiological carrier that might oxidize them in intact cells. Eventually, in the nineteen-fifties, a flavoprotein was obtained capable both of oxidizing $NADH_2$ and of being itself reoxidized by cytochrome c. On the face of it, this could fulfil the role of X in the scheme on p. 78; but there are serious objections to the view that it plays a major physiological role. Whether or not it ultimately turns out that flavoproteins of some kind act between NAD and cytochromes in a major respiratory chain (either with or without the participation of other types of carrier), the story of all these uncertainties shows the difficulties inherent in attempts to build up a picture of metabolic machinery and its functioning merely by studying its parts after isolation.

What grounds are there, then, for confidence that such a picture can ever be built on convincing foundations? Perhaps one can point to two main ways of tackling the general problem. One is investigation of events at levels of organization intermediate between intact organisms and molecules in free solution; the other relies on the use of isotopes. Both have come into prominence only since the Second World War. Both profoundly affect contemporary biochemical thought and practice, and it is appropriate to discuss here how they can be used to meet the objections to the "analytical" approach. A few words will be said, too, about a third type of technique that is also relevant, though of much more restricted applicability; this involves observing intact cells and tissues spectroscopically.

Meeting the objections (i) Intermediate levels to bridge the gap

The disparity in degree of organization between intact organisms and purified substances is great; much has to be destroyed in passing from one to the other. But the gulf between them no longer yawns as impassably as it used to (p. 109), because studies can now be made at a whole series of intermediate levels of organization.

The organization of living beings falls into a series of levels arranged in the form of a hierarchy, as in organizations such as the civil service or the army, each level embracing the members of lower levels in characteristic "organizing relations". For the present purpose, five main steps in the hierarchy may be distinguished (cf. also p. 110). After the whole organism come organs such as liver and kidney; then come tissues and cells, then subcellular particles, and finally molecules free in solution.

Biochemical processes can be investigated with preparations representing any of these steps. To do biochemical work, it is not necessary always to destroy all levels of organization above the molecular. Quite on the contrary, it is necessary sometimes to proceed less violently—just as, to find out about the inner workings of a washing-machine, it would be necessary to lay aside the sledge-hammer and dismantle the machinery gently stage by stage.

In the case of intact organisms, biochemical study might take the form of feeding experiments—analysing the diet, or adding special substances to it, and finding out what is excreted. At the organ level, the corresponding experiments can be done by perfusing an excised organ with blood or with an isotonic salt solution; test substances can be added to this solution, and the emerging liquid analysed to see what transformations have taken place. Tissue activity can be studied by suspending thin slices in an appropriate salt solution; the cells in such slices metabolize for some time in a manner approaching the normal. Micro-organisms offer the advantage, as experimental material, that they can be grown as homogeneous cell populations, with no level of organization above the cellular.

For studies at the subcellular level, cells can be broken by homogenization. The way in which subcellular particles can be separated by differential centrifuging, and the precautions that must be observed to keep them reasonably intact functionally, have already been described (p. 72). Finally, the individual units of catalytic machinery, enzymes and their cofactors, can be obtained in free solution. Some can simply be extracted by aqueous media; some are rather firmly bound to structures such as the mitochondrial membranes, and special treatments have to be applied to solubilize them. Once in free solution, the methods for separating enzymes from each other, so as to obtain homogeneous collections of molecules, are the methods for protein purification (Chap. II).

The levels of the hierarchy form a continuous series with an impressive degree of coherence. There are no abrupt breaks—no gaps of more than about one order of magnitude. Very roughly, 10^{-7} cm (10 Å) can be taken as representative of the molecular order of magnitude; the dimensions of subcellular particles are in the region 10^{-6} to 10^{-4} cm; cells at around 10^{-3} cm (10 μ) and larger are not much smaller than macroscopically visible structures on the millimetre scale. (A qualitative discontinuity in methods of study is, however, worth noting. One individual animal at a time is dissected, a single liver is perfused; but cells and molecules are normally handled in large collections. Isolating a substance means getting not a single molecule, but a homogeneous collection of molecules. The results obtained with such a collection are automatically subjected to statistical smoothing; so that it is not surprising that biochemical reactions with purified substances are in general more accurately reproducible than experiments on whole animals or large parts of them. A rough comparison of the statistical status of work at different levels is revealing. A solution of a substance as dilute as one-thousandth molar contains about 10^{18} molecules of it in 10 ml; a bacterial culture may grow to something like 10^9 cells in the same volume; this is the order of magnitude of the total number of human individuals available on the earth for any experiment a super-scientist might want to do on them.)

The continuity of the hierarchy offers hope that by systematic, stage by stage comparisons the gap between molecules and large animals can be bridged, data at the molecular and whole organism levels being brought into valid correlation. The task of bridging across intermediate levels is perhaps the major challenge facing contemporary biochemistry. It forms a sort of *leitmotif* running through the remainder of this book, appearing in the form of attempts to tackle the important general problem in various specific contexts. Thus it shows itself as the effort to relate vitamin activity for whole animals to cofactor activity for enzymes (Chap. VI). In the case of chemotherapy, it takes the rather similar form of relating the clinical activity of drugs to their chemical properties (Chap. VII). For muscle, there are the ingenious attempts to picture the molecular events that underlie visible contraction (Chap. VIII). Last, but far from least, there is the large problem of how the molecular structure of nucleic acids can determine macroscopic hereditary characters (Chap. IX).

So far in the present chapter, specific examples to illustrate the general discussion have been drawn from the study of biological oxidations. In this field, early successes were obtained at high levels of organization, and subsequent advances have been made largely by extensions to progressively lower levels. It is easy to see how increase in understanding has depended on studying systems of varying degrees of disruption. Lavoisier, working with whole animals, was able to characterize the overall chemical result of respiration (p. 91); the extent of his ignorance of its details is indicated by his suggestion that the oxidation of carbonaceous material to carbon dioxide occurs in the lungs. During the nineteenth century, analysis of blood for its contained gases made it possible to follow the course of oxygen via the blood to the rest of the body For some time, it was generally believed that oxidation takes place in the blood itself as it passes through the capillaries; not until the eighteen-seventies was it finally established that the tissues themselves are responsible for nearly all the oxidation.

Attempts to extend the study of the processes involved to the subcellular level early in this century soon ran up against particles. Aqueous tissue extracts, though they contain the enzymes for anaerobic breakdown of carbohydrate (Chap. III), show little of the oxygen uptake of intact tissue, which is effected mostly by catalytic systems bound to particles, the mitochondria (p. 75). Some of the catalysts are, indeed, so firmly bound that they are difficult to obtain in true solution and have been studied largely in broken particle preparations (notably cytochrome *a*, p. 79). Other enzymes and cofactors were obtained in soluble form and purified; this has been an active field of research since the period between the two World Wars.

Before leaving the subject of systems of intermediate degrees of complexity, it is worth noting that such systems can be reached from below as well as above—by partial reconstitution as well as by partial degradation. Thus, on the one hand, enzyme reactions are being studied not only singly but also in the groups which are catalysed by groups of enzymes in various particle preparations (notably, in the case of oxidation reactions, fragments of mitochondrial membranes obtained by breaking mitochondria in not too drastic ways). On the other hand, isolated enzyme and particle preparations can also be mixed to study their mutual interactions, in the hope of elucidating those of physiological significance.

Such reconstituted systems are, in a sense, models which aim to copy biological effects, but they differ in an important respect from the models discussed earlier, where the constituents are non-biological. Reconstitution is a natural complement to analysis, not an attempt to by-pass it. Putting together, indeed, follows on from taking to pieces with the same logical and historical inevitability that made synthetic follow analytical organic chemistry.

Reconstitution experiments may, perhaps be looked on as steps towards a "synthesis of life"—early, unsteady steps, but a sounder beginning than the leaps into the void represented by models made from materials obviously different from the biological ones. Attempts at synthesis are not likely to be profitable before knowledge of constitution is rather detailed. The artificial homunculus in the test tube is to biochemistry rather as alchemical gold has been to chemistry; without insisting on its ultimate impossibility, one can confidently predict that premature attempts at short cuts by producing superficial imitations will not get nearer the mark than the longer road of analysis and its complement, reconstitution.

Meeting the objections (ii) *Spectroscopic reconnaissance*

Before discussing other ways of meeting the objections to the analytical approach, it may be as well to restate the general problem at issue. If biochemistry involves disrupting organisms, how can one determine which of its results apply to organisms before disruption? Seen in a broader context, this question is only one aspect of the spectre that haunts any form of experimental biology—the suspicion that what is being studied is not a "natural" event but an artifact of the experimental procedure. Experiment implies active interference; the very fact of this interference may crucially affect the interpretation of the results.

To indicate how general the issue is, one may point to the analogous situation that arises in the context of microscopy. Much of the data of microscopy could not be obtained except by the use of fixed and stained preparations. How can one be sure that what one sees is not an artifact produced by the fixing and staining procedures? It was believed, at one time, that protoplasm has a network structure on the scale visible by ordinary light microscopy—until it was shown, at the end of the nineteenth century, that such networks are readily formed by coagulating protein solutions with reagents of the sort used as fixatives. Because of observations such as these, microscopy has not been without its nihilist faction, voting for no confidence in any of its conclusions. Clearly, such extremism is out of place; what is called for is not out-and-out scepticism but a thoroughly cautious approach, combining the examination of fixed and stained preparations with such studies as can be carried out on living material and with critical investigation of the changes involved in fixing and staining.

In the biochemical context, the problem is that of finding out about the reactions really occurring in intact organisms, as distinct from those that can be studied in systems of varying degrees of disruption. Further, there is also the problem of deciding how these reactions fit into metabolic pathways.

In isolated instances, reactions can be studied in intact cells spectroscopically. Some cell materials are coloured and have characteristic absorption spectra which change when they react. Light passing through the cells can therefore bring information about their conversion from one chemical state into another.

Probably the most notable case is that of the cytochromes (p. 78) which, as their name implies, are cell pigments and absorb light in the visible range. In the reduced form containing ferrous iron, they show sharp absorption bands at characteristic wavelengths. When oxidized to the ferric form, the sharp bands disappear and are replaced by more generalized absorption. This makes it easy to follow the oxidation and reduction of cytochromes in their natural habitat by observing them with a spectroscope. A simple spectroscope merely breaks up light into its constituent wavelengths, so that on looking through it one sees the colours of the spectrum arranged in order from blue to red. If a pigment with sharp absorption bands is placed in front of the spectroscope, black stripes appear in the spectrum at those wavelengths at which the pigment absorbs light. Reduced cytochromes give rise to such black bands, but the more generalized absorption of oxidized cytochromes only leads to diffuse shading which is not strikingly obvious to the eye. Appearance of the bands therefore indicates that the cytochromes are undergoing reduction, and disappearance indicates oxidation.

It was by this simple technique that the cytochromes were discovered and their behaviour studied by Keilin in Cambridge in 1925. (Actually, the discovery was a rediscovery, for the pigments had been observed many years before, but they had been written off as artifacts derived from haemoglobin, which they somewhat resemble chemically. Biochemists have to be on their guard against artifacts, but the critical attitude can be taken too far.) Keilin found cytochromes in a great variety of animals, micro-organisms and plants. In baker's yeast, oxidation and reduction of the pigments was easy to observe. An aqueous suspension of the yeast showed the bands of the reduced forms; when air was rapidly bubbled through the suspension, the cytochromes became oxidized and the bands disappeared; when the current of air was stopped, or when it was replaced by one of nitrogen, the bands reappeared.

It was even possible to watch cytochromes acting in an intact animal and to see that muscular exertion (leading as it does to increased breakdown of respiratory substrates and hence increased flow of hydrogen or electrons along respiratory chains) brings about cytochrome reduction. "To understand the function of this pigment [i.e., group of pigments] it is important to find out in what state this pigment is present in a normal living organism", wrote Keilin in his 1925 paper. "The main difficulty in answering this question consisted in finding suitable material for such an investigation. One insect, however, the common wax-moth (*Galleria mellonella*) answered the purpose. Several active specimens of this moth were selected out of a large stock bred in the Laboratory, and the thorax of each was carefully cleaned from scales. These specimens were then attached by the ventral surface to a slide (by means of small droplets of gum arabic) and the thorax carefully examined with Zeiss microspectroscope and a strong light. (In this condition the insects remained alive for long periods, and even after three hours the females, in spite of being attached to the slide, went on ovipositing.)

The following are the results of these observations:—

(1) Female of *Galleria* remained very quiet, except for occasional expulsion of an egg and a somewhat rhythmic movement with ovipositor. The thorax being of a yellow colour showed better the long-wave portion of the spectrum, but no absorption band could be detected.
(2) In specimens of males and females which began to struggle constantly, vibrating their wings in efforts to detach themselves from the slides, the bands of cytochrome gradually appeared . . .
(3) When these specimens ceased to move and stopped the vibration of the wings, the bands became very faint and hardly detectable . . .

The absorption bands of cytochrome shown by *Galleria* after exertion . . . are never so strong as they appear in specimens exposed to pure N_2 or to the vapours of KCN. [Cyanide in low concentrations stops the oxidation of cytochromes.] This fact indicates that in natural conditions cytochrome is in the oxidized form, and that during exertion, however great, cytochrome becomes only partially reduced.

The above experiments with *Galleria* and the previous observations on yeast show that cytochrome acts as a respiratory catalyst . . . In the living organism the state of cytochrome as seen spectroscopically denotes only the difference between the rates of its oxidation and reduction."

In this investigation, it can be seen, an elegant conception was put into operation with means of almost ludicrous simplicity. Much more refined apparatus for this type of study has since been developed; the procedure can be made quantitative and more sensitive, and its scope extended beyond the range of visible light to the near ultra-violet region. Besides cytochromes, flavoproteins can be studied in the visible range and NAD in the near ultra-violet.

Nevertheless, it cannot be said that the spectroscopic approach is of anything more than very limited applicability. It might be compared to aerial reconnaissance, in that it can detect only relatively gross changes fairly easily made visible. It can be used only for cell components whose functioning is accompanied by spectral changes at characteristic wavelengths where absorption by other cell components is low. In regions of the spectrum where absorption is high, any but the most drastic changes are so small by comparison as to be effectively masked or "camouflaged"; this applies in general to the infra-red and far ultra-violet ranges.

Meeting the objections (iii) Isotopic espionage

To penetrate the camouflage, it is necessary to resort to espionage, sending spies into the cells to report on what cannot be seen from outside. The biochemical spies are isotopic tracers. It was a crucial development in biochemistry when isotopic methods were introduced around the time of the Second World War, for they provided the first reasonably general way of following chemical changes without disrupting organization.

The concept of isotopes is now widely familiar. Chemically the same because of their identical electronic make-ups, two isotopes of an element differ in that their atoms have nuclei of different mass. The main natural isotope of carbon, for instance, is ^{12}C, meaning that its nucleus has a mass twelve times that of a hydrogen atom. The nucleus of ^{14}C has a mass fourteen times that of a hydrogen atom, but at the same time it has the same complement of six electrons as ^{12}C, so that it forms the same series of chemical compounds.

Isotopes can be distinguished by physical properties. In the case of ^{14}C, detection and estimation are easy because this isotope is radioactive—the nucleus decomposes, giving off β-particles which can be measured in terms of counts per minute in a radioactivity counter. The same applies to ^{32}P, a radioactive isotope which, because of the importance of phosphate compounds (p. 60), is very useful in biochemical research. (Normal phosphorus is ^{31}P.) In the case of stable, non-radioactive isotopes, estimation depends on the difference of mass. The most important stable isotopes used in biochemistry are ^{2}H, heavy hydrogen or deuterium, with twice the mass of ordinary hydrogen; and ^{15}N, heavier by one unit than the most abundant natural isotope, ^{14}N.

Being chemically the same but physically distinguishable, isotopes can be used as tracers to trace metabolic pathways. The abnormal isotope can be administered in the form of an isotopically labelled compound, which may be a normal metabolite except in the isotopic respect. Later, by finding out in what substances the isotope is now present, the metabolic transformations undergone by the original compound can be detected. The isotope label thus traces the paths from precursors to products. If the spectroscopic approach depends on cells or tissues being in a literal sense transparent to light, isotopes make them biochemically "transparent" in a metaphorical sense. The important point is that, in both cases, intact metabolic machinery can be used to bring about the transformations under study. It may be necessary to disrupt the system at the end of the experiment to find where the isotope has got to, but the processes by which it got there went on under physiological conditions.

Isotopes can thus discharge their function of biochemical espionage in a rather subtle way. Introduced into organisms under the cloak of their normal electronic shells, the abnormal nuclei mingle unobtrusively with the chemical population and spy out chemical activities without disturbing them by disrupting the natural organization of the machinery that brings them about. Masquerading as normal metabolites, the isotopically labelled molecules get into normal metabolic pathways and set the abnormal atoms as markers along it.

The isotopic tracer method is applicable to most fields of intermediary metabolism. In the particular case of the respiratory chains, unfortunately, its usefulness is very limited. The reasons for this are quite easy to see. The material passed along respiratory chains consists of hydrogen atoms or electrons. Electrons are impossible to label; hydrogen atoms can be effectively labelled only if carbon-bound, for if bound to oxygen or nitrogen they normally exchange very rapidly with the hydrogen atoms of the aqueous medium. On much of its route along the respiratory chains, hydrogen seems to be bound in the exchangeable way, so that deuterium cannot be used to track its path. It seems safe to say that, but for this limitation of isotopic tracer technique, the considerable present uncertainty about the intermediate carriers and reactions of respiratory chains (p. 78 and p. 94) would not persist. Metabolic pathways of comparable importance involving carbon, nitrogen or phosphorus atoms have by now been mapped, at least in outline, by the use of suitable isotopes.

As an example of the successful use of isotopic tracers, it seems appropriate to take photosynthesis, in a way the complement of biological oxidation.

Isotopic carbon in the study of photosynthesis

The process by which green plants in the light build up organic materials has long been recognized as one of fundamental importance, since both the plant and animal kingdoms depend on it. Until quite recently, however, it was impossible to trace the intermediate steps by which the carbon atoms of carbon dioxide become those of carbohydrate. Up to the time of the Second World War, discussion was still going on about the old formaldehyde hypothesis put forward by Baeyer in 1870. This was based on the crude model system in which formaldehyde, HCHO, polymerizes under the influence of alkali to give carbohydrate. Attempts to implicate formaldehyde as an intermediate in photosynthesis never brought conviction, and it is now known that plants use a very different route to make carbohydrate.

It was during the nineteen-forties that the two techniques became available which made it possible to map the path of carbon in photosynthesis. One was isotopic tracer technique using the radioactive isotope ^{14}C. The other was paper chromatography (p. 27); in the pre-chromatography days, separation of the many metabolic intermediates in photosynthesizing systems would have been a gargantuan task.

The successful combination of these two techniques was largely the work of Calvin and his co-workers in California during the early nineteen-fifties (Calvin won the Nobel prize for chemistry in 1961). Their work on photosynthesis illustrates in a particularly elegant way the value of isotopic tracers in the elucidation of metabolic pathways. Thus it was not dependent on the preparation of cell-free extracts capable of performing the process—differing in this notable respect from the classical work on the anaerobic breakdown of carbohydrate (Chap. III). ^{14}C was administered in the form of $^{14}CO_2$ to intact plants; the reactions being studied were *in vivo*, not *in vitro* ones. Furthermore, many of the reactions involved in carbohydrate synthesis are the steps of carbohydrate breakdown in reverse, and study of isolated enzyme systems alone could hardly have given convincing evidence for the participation of any particular step in either pathway. "In order to establish the pathway of carbon reduction during photosynthesis", Calvin has written, "a method was required for determining *in vivo* not only the reactions, if any, which are unique to photosynthesis, but also which of the many enzymatic transformations of carbon compounds already known are important in this process. In addition it was necessary to discover not only individual steps but also to place these steps in their correct relationship to one another. In view of the dual role of most of these enzymatic systems (in respiration and in photosynthesis) it would be impossible to identify photosynthetic steps and establish their inter-relationships from a knowledge of *in vitro* systems only".

TWO APPROACHES TO BIOLOGICAL EXPLANATION

In the experiments of Calvin and his collaborators, then, $^{14}CO_2$ was supplied to intact, actively photosynthesizing plants. Unicellular green algae like *Chlorella* and *Scenedesmus* were used most, since with them it is easier to obtain uniformity with respect to plant material, illumination and nutrition, though experiments were also done with the leaves of higher plants. After a suitable time interval, metabolic activity was stopped by running the algal suspension into four times its volume of boiling absolute alcohol. Extracted material was analysed by paper chromatography. Radioactive spots on the paper after chromatography were located by "radioautography". This means placing the paper in the dark in contact with a sheet of X-ray film; opposite each radioactive spot, the film becomes exposed and shows a darkened area after development.

Within a short time—a minute or so—of $^{14}CO_2$ assimilation, the ^{14}C label appeared in many different compounds, as indicated by the formation of many radioactive spots on paper chromatograms. Clearly, many reactions occur rapidly. The chemical complications, therefore, soon become considerable, but—CO_2 being such a simple compound—the initial steps are easy enough to follow; they illustrate quite adequately the general nature of the approach.

To identify the initial reaction of carbon fixation, times of exposure to $^{14}CO_2$ were cut down to restrict labelling to the first product formed. With periods of exposure of as little as five seconds, the predominant labelled compound was 3-phosphoglycerate. The primary intermediate of carbon assimilation thus turned out to be a compound already well known in biochemistry, for it lies on the major route of sugar breakdown (Fig. 11). It is, of course, a three-carbon compound, and it seemed at first as though carbon dioxide fixation involves condensation with a two-carbon compound. Actually, however, the carbon dioxide acceptor was identified as the five-carbon compound ribulose-1,5-diphosphate; combination with carbon dioxide is followed by a split which leaves two molecules of phosphoglycerate.

$$\begin{array}{c} CH_2O\text{\textcircled{P}} \\ | \\ CO \\ | \\ HCOH \\ | \\ HCOH \\ | \\ CH_2O\text{\textcircled{P}} \end{array} + CO_2 + H_2O \rightarrow \begin{array}{c} CH_2O\text{\textcircled{P}} \\ | \\ CHOH \\ | \\ COO^- \\ COO^- \\ | \\ CHOH \\ | \\ CH_2O\text{\textcircled{P}} \end{array} + 2H^+$$

Ribulose-1,5-diphosphate   3-Phosphoglycerate (2 molecules)

THE BIOCHEMICAL APPROACH TO LIFE

(Ribulose is a ketose sugar, containing the grouping $CH_2OH \cdot CO-$, like fructose (Fig. 11), from which it differs in having one carbon atom less, being a pentose rather than a hexose.)

The conversion of phosphoglycerate into glucose and the glycogen-like polysaccharide starch occurs by a route that is substantially the reversal of that taken during breakdown, shown in Fig. 11. (Steps 1 and 3, however, are not reversible exactly as shown; although fructose-6-phosphate can be formed from fructose-1,6-diphosphate, and glucose form glucose-6-phosphate, no ATP is formed in these reactions—cf. p. 150.)

Only a small part of the phosphoglycerate formed is available for building up into starch and other end-products of photosynthesis. To maintain a supply of ribulose diphosphate and thus keep the process going indefinitely, most of the phosphoglycerate must be used to regenerate the carbon dioxide acceptor. Of every six molecules of the three-carbon compound, one is used to build up starch and the other five are used to form three molecules of the five-carbon compound. The steps by which the latter process takes place have been elucidated, but they are fairly complex and need not be detailed here. It is enough to see that the carbon balance is like this:—

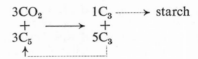

Whether the phosphoglycerate is to regenerate ribulose diphosphate or to form starch, the first two steps involved are the same—steps 6 and 7 of Fig. 11 in reverse, yielding in turn 1,3-diphosphoglyceric acid and glyceraldehyde-3-phosphate. Besides the catalytic apparatus of the relevant enzymes, therefore, two things have to be supplied—ATP and reducing power. The ATP is required to phosphorylate the carboxyl group of phosphoglycerate, the reducing power to reduce the carboxyl-phosphate to an aldehyde group. (In photosynthetic systems, this reducing power seems to be supplied in the form of $NADPH_2$ rather than the $NADH_2$ shown in Fig. 11.) To provide ATP and reducing power is the function of the light-dependent reactions of photosynthesis. Light provides the energy for ATP synthesis, and is also used to effect—in some indirect way about which much remains to be learned—the splitting of water to give the hydrogen used to effect reduction and the oxygen which is evolved.

TWO APPROACHES TO BIOLOGICAL EXPLANATION

A dated controversy—"mechanism" versus "vitalism". The organization hierarchy

The old question of "mechanism" versus "vitalism" is still dutifully trotted out in the first or last chapters of a large number of books on various aspects of biology—an amazingly large number, for in the context of modern thought it requires a degree of historical imagination to understand the importance this issue assumed. Why it aroused a good deal of passion is easy enough to see, for mutual misunderstanding has always been good at that; but it is harder to appreciate how it came for so long to be regarded as the central issue of biological theory.

Can biology be explained in terms of physics and chemistry, or must some other, "higher" agency be postulated to account for the phenomena of life? This, or something similar, is a form in which the question was often put up for debate. As it happens, definitions of the subject-matter of physics and chemistry do not normally contain any explicit limitation to non-living material and forces. Pedantically, therefore, the question can be denied meaning anyway. On a common-sense level, however, its implication is clear enough. Are living and non-living things governed by the same basic laws or not? Are they fundamentally similar or fundamentally different?

The obvious answer is that they are similar in some ways and different in others. Whether the similarities or the differences are to be rated more significant is an issue of something less than crucial importance. It hardly seems meaty enough to account for all the fuss that went on during sizable portions of the nineteenth and twentieth centuries (though it might have made an acceptable main course for a disputation among mediaeval scholastics).

The whole debate becomes largely superfluous if the problem is viewed in terms of living things as organisms—complex entities made up of parts organized in a hierarchical series of levels of organization (p. 95). Parts organized into a whole organism do not behave in the same way as parts separate or disorganized; and different modes of organization can lead to different behaviour. Now living things are organized in characteristic ways, as is obvious even to the most superficial observer. Inevitably, therefore, they behave in characteristic ways. Their parts may be just the same as the parts of other things (and it has long been agreed that they are); but the laws governing the parts do not give the laws governing the organized whole. Qualitatively new kinds of phenomena can emerge at higher levels of organization. A motor-car can apparently violate the physical law of gravity by climbing uphill. None of its parts can, nor can all its parts together unless properly organized—the organized whole behaves differently from its parts isolated or disorganized.

For this reason, it is absurd to suppose that all the laws of life can be fished out of solutions—liquids containing no level of organization above the molecular. Work on component parts is essential in studying organized wholes—otherwise, understanding can be no more than superficial in the most literal sense of the word; but it is not enough to study only the parts themselves, to the exclusion of the organizing relations between them. To find out about these specific spatial and functional relations, which turn a collection of components into an organism, the parts, the whole and intermediate levels of organization must all be investigated. It is the hierarchy which reconciles the claims of the analytical approach with the rejoinder that intact organisms are the biological reality to be explained. The organization hierarchy, forming as it does a bridge between parts and whole, is thus one of the really vital, central concepts of biology.

Mechanism and vitalism each took many forms—they were highly personal faiths which appeared in almost as many different manifestations as there were people who thought about them. This is not the place to attempt a complete *exposé* of all the heads of these many-headed monsters. There are three points, however, which deserve mention because of their relevance to topics already treated in this chapter. The first concerns the adequacy of models as a means of explanation, the second the continuity of the organization hierarchy and the third the importance of not neglecting levels of organization above the molecular.

The weaknesses of biological explanation by means of non-biological models have been pointed out (p. 90); but in the context of the mechanism-vitalism controversy there was a limited sense in which a biological effect could be said to be "explained" by imitating it with a non-living model. Vitalists were at their most unfortunate, perhaps, in those excursions in which they mixed metaphysics with science, postulating vital principles under various names whose intervention was assumed because inanimate forces did not seem adequate to explain some biological phenomena. While this was the point at issue, it was quite relevant to attempt to simulate the phenomena with non-living systems; to show how a process *could* happen in living organisms was enough to refute this vitalist argument from inadequacy. When mechanists managed to do this, however, they were not always careful enough to avoid the pitfalls of success. The dangers of thinking in terms of models are, firstly, that the models tend to become invested with an aura of biological reality to which they have no claim; and, secondly, that energies tend to be diverted into the study of models instead of the direct study of living organisms and their parts. Vitalists were in a much stronger position when they issued valuable reminders that the proper study of biology is life—how it *does* work, not how it *could*.

The threat of divorce between chemistry and biology would never have become as serious as it did had the continuity of the organization hierarchy always stood out as clearly as it does to-day. By the beginning of the present century, biological analysis proceeded down to the level of cells; these were regarded as the units of life (and the cellular level of organization is, of course, a particularly important one for biology). Chemistry dealt with atoms and small molecules. Between the entities with which the two sciences typically concerned themselves lay an apparently ominous gap of several orders of magnitude, which could be bridged only in a very unsatisfactory way by vague talk of the "colloidal state of protoplasm". More recently, however, a pincer movement has converged from above and below on this gap. From above have come increases in microscopic resolution, for instance, notably by electron microscopy; from below, chemical methods have taken larger and larger molecules more effectively into their grasp. It is no accident that biochemical interest is now focussed specially on that critical region of the hierarchy where macromolecules merge into subcellular particles (Chaps. II and IV; pp. 162, 176).

Model systems of chemical interest normally take the form of reactions proceeding in free solution or the gas phase—in general, in systems with no level of organization above the molecular. Because of this, mechanism in its chemical aspect was often criticized for ignoring the higher levels—for seeing in living things only a mass of substances and no architecture. Insistence on the importance of this architecture was another valuable reminder that came from vitalists. Though it led some of them into an obstructionist resistance to analysis, others made timely contributions by keeping the higher levels of organization in the public eye.

In many vitalist attacks on the application of chemical methods to living things, however, a fundamental flaw was to deny the chemical approach any stake in levels above the molecular. It was taken to apply to "disorganized" systems only. If this were so, it would not apply to man-made machines any more than to intact living organisms. The absurd position would be reached of denying the applicability of chemistry to motor-cars because they differ from any of the substances composing them in being able to climb uphill. ("Mechanism" in its physical sense clearly implies organization; it is paradoxical that in the chemical context it should have been taken to deny organization.)

There is no point in arbitrarily splitting the organization hierarchy into precisely delimited spheres of influence belonging to various sciences. The hierarchy is an extensive one—it can usefully be considered to go as far down as atoms and subatomic particles and as far up as societies of living individuals. It thus extends further than the range with which any single branch of science normally wants to concern itself. But to attempt, on grounds such as this, to restrict particular sciences to particular ranges of the hierarchy is misguided. Although a science of itself finds its own characteristic range, there is no need to imprison it there. Chemistry, for instance, characteristically operates at the levels of atoms and molecules; but there is considerable profit, and no logical inconsistency, in allowing it to penetrate below the atomic level to arm itself with electronic theories of valency and reaction mechanisms, and above the molecular level to learn from experience in chemical factories or living organisms.

Biochemistry does not ignore organization, as it has been accused of doing—it merely selects a certain range of levels as its normal and proper subject-matter. In general, it chooses to take the atomic and subatomic levels as given. The molecular level is its own special and characteristic concern. This does not mean, however, that it must neglect higher levels; quite on the contrary, it must take account of all levels up to that of whole organisms. The study of life in molecular terms embraces the relations between molecules just as much as the properties of the molecules in themselves.

An artificially imposed ceiling for biochemistry at the molecular level would be absurd—it would take away the major part of its proper subject-matter. To bring out the full absurdity of such a ceiling, it may well be compared to one for chemistry set at the atomic level. Chemistry did not get off to a sound start until it learned to concentrate on atoms, and atoms remain its main explanatory entities. It might conceivably be argued that it should give up studying compounds and concentrate on so exhaustive a study of atoms that the ways in which they enter into combinations, and hence the chemistry of compounds, would become deducible from the laws that emerged. Two flaws in this argument immediately spring to mind. Firstly, even the most thorough understanding of the laws of chemical combination between atoms, though it might make possible the prediction of what compounds *could* exist, could not by itself tell what is actually the structure of any given compound. Secondly, what knowledge we have of the laws of chemical combination has in fact been got largely by studying the compounds that atoms form, and could hardly have been got any other way.

The respective positions of atoms and molecules in chemistry correspond in significant ways to the positions of molecules and organisms in biochemistry. The organization of molecules into living organisms is just as much part of biochemistry as the organization of atoms into molecules is part of chemistry.

VI

The High Road and the Low Road—Vitamins and Coenzymes

Two roads to a dramatic meeting

ALL the talk in the last chapter about organization hierarchies and correlations between different levels of them may seem rather abstract and philosophical; but the story of vitamins and coenzymes brings out its very real relevance to the strategy of biochemistry at work. This story is a case-history, illustrating how a sweepingly general current of biological thought works in the background, unobtrusively but inexorably shaping the pattern of discovery. The relation that has been uncovered between the two apparently quite distinct concepts—vitamins on the one hand, coenzymes on the other—and the attempts to consolidate the contact that has been achieved between knowledge in the two fields, show poignantly both how desirable and how difficult it is properly to correlate observations made at different levels of organization, from the whole animal to molecules. Here, then, is an expression in particular and practical terms of a major dilemma and challenge of biological theory.

Few adults nowadays have no idea what vitamins are; it is a subject which, after all, has been written about *ad nauseam* at all levels of scientific sophistication from technical jargon to journalese. Not all the current ideas are accurate, however. Many think of vitamins first in terms of pills to be bought as a kind of drug, in the hope—usually over-optimistic—that they will make children grow, cure colds or generally provide "pep". "How did people manage before vitamins were discovered?" is a question that has not gone unasked. In fact, of course, they are normal constituents of natural diets, present in small amounts but vital in importance.

Many vitamins are known to form part of coenzymes. Coenzymes are small but essential items of the catalytic equipment of cells; not being proteins, they are not enzymes themselves, but they are necessary for the functioning of certain important enzymes. Like other body constituents, coenzymes need periodic replacement. In those cases where a part of a coenzyme molecule cannot be made in the body from protein, carbohydrate and fat, it must be taken in as an additional component of the diet, and it then counts as a vitamin. Vitamins, in fact, are otherwise irreplaceable spare parts for the body's chemical machinery.

Like characters in a romantic novel, vitamins and coenzymes started their careers as strangers to each other, unaware of the close blood-relationship invisibly binding them to each other. In the event, they have recognized each other as half-brothers, both begotten by biochemistry—the one out of intact organisms, the other out of enzyme extracts. The plot of the novel falls into two main parts. The first part traces the histories of the two characters up to the point where their paths crossed and intertwined—emphasizing, for the sake of dramatic effect, how different were the routes along which they had previously developed. One had taken a road along a high level of organization; vitamins were discovered by investigating what intact animals require in their food to maintain health and life. The other had taken a road along a low level of organization; coenzymes were discovered by investigating what extracted enzymes require in the medium to show catalytic activity. With the meeting and mutual recognition of the two characters—an incident engineered as if by accident, but really made inevitable by the fundamental relation between them—the story reaches its turning point; it then deals with their attempts to achieve satisfactory communication and understanding between each other, and with the valuable help each has on occasion been able to render the other.

The nutritional road—vitamins (i) Deficiency diseases

Two strands of evidence, both nutritional, converged early in the present century to make possible the discovery of vitamins. One concerned deficiency diseases, the other arose out of the analysis of diet.

Empirical knowledge of deficiency diseases can—with hindsight—be traced back many centuries. A clinically accurate account of beri-beri, including not only the symptoms but also foods that cure it, has been found in the Chinese literature of the fourteenth century. Many of the early European descriptions are of scurvy, with its swollen, spongy gums and haemorrhages under the skin due to abnormal fragility of the blood capillaries. Striking accounts were left by the seamen of the Age of Discovery, who were often at sea for months without fresh food, eating a preserved, salted diet. Mortality amongst them was heavy and continued for centuries, for although the curative effect of foods like oranges and lemons was noted repeatedly, the observations remained scattered and incidental. As early as 1601, Sir James Lancaster introduced oranges, lemons and lemon juice into the diet on ships of the East India Company. Captain Lind, an English naval physician, published in 1757 a *Treatise on Scurvy* in which he described how even severe cases could be cured within a week by "salads, summer fruits" and similar foods.

It hardly occurred to people at the time that fruit might be supplying an essential nutritional factor. Most theories of scurvy attributed the disease to excess of salt, and fruit was regarded as supplying an antidote rather than remedying a deficiency. As for guessing at the nature of the active principle, it was understandable that it was the acidity of citrous fruits that was seized on. This may help to explain why Joseph Priestley advocated the use of soda water; his interest in carbon dioxide (or "fixed air") had been stimulated by the circumstance of living near a brewery in Leeds, and he published a process for making soda water in 1772. The British Navy, however—perhaps as much by luck as judgment—stuck to lime juice, the widespread use of which from the end of the eighteenth century onwards kept it relatively free from scurvy and was an important factor in its success.

The possibility of curing beri-beri by a change in diet was demonstrated by another naval physician, Admiral Takaki. During the eighteen-eighties, he virtually eliminated the disease from the Japanese Navy by increasing the allowances of vegetables, fish and meat in a diet which otherwise consisted largely of rice. He attributed his success, however, not to a new accessory food factor but to higher protein intake. (There is more than coincidence in the way that naval medicine dominates the early history of deficiency diseases. It was the restricted nature of ships' diets that set up the nutritional stresses which raised the problem in its acutest form and threw down the sharpest challenge to find a solution.)

The climate of opinion late in the nineteenth century was not favourable for the idea to take root that some diseases could be due to deficiency of minute amounts of food factors. Since the work of Pasteur, Koch and Lister in the eighteen-seventies, the germ theory of disease had been sweeping all before it. When one disease after another was being attributed to specific micro-organisms, the world of science was naturally loth to accept the view that some diseases are due to lack of substances whose nature could not even be guessed at, in amounts that could not be measured. The concept of deficiency diseases was established only when it became possible to induce such conditions in experimental animals. This was first done in the eighteen-nineties by Eijkmann, in what were then the Dutch East Indies. Feeding hens on rice that had been milled and thereby deprived of its bran, he noted that they developed nervous symptoms which he recognized to be like those of beri-beri in man. These symptoms could be cured by adding rice polishings to the diet. Eijkmann extracted the factor responsible for the cure from rice polishings, partially purified it and found it to be water-soluble and dialysable (p. 118). He himself thought of it as acting as an antidote against some other disease-producing factor, but his colleague Grijns in 1901 clearly formulated the idea that it is a dietary constituent other than protein, carbohydrate, fat or salt which is itself indispensable to health.

By 1912 the time was ripe for systematization of the knowledge that had been collected. This was done notably by Funk, a Pole working at the Lister Institute in London, in his theory of "vitamines" (so-called in the mistaken belief that they are all amines—most are not). Funk collected the evidence for the dietary origin of four diseases. Two of these were beri-beri and scurvy, due respectively to lack of the substances we now call vitamin B_1 or thiamine, and vitamin C or ascorbate. The third was pellagra, the disease characterized by the "four d's"—dermatitis, diarrhoea, dementia and eventually death—which was prevalent up to the nineteen-thirties in the south-east of the United States of America. It is now attributed primarily (though not without complicating factors) to shortage of nicotinate, a member of the B group of vitamins. The fourth disease was the bone deformity known as rickets, and here Funk had to recognize that the evidence was conflicting. Some of it pointed to an association with lack of "good" fat in the diet—"good" fat in this context meaning animal fat. (Fashions in dietetics are fickle, however; nowadays, in the different context of cardiovascular disease, we are told that vegetable fat may be preferable.) Rickets is now known to be due to lack of vitamin D, in which some animal fats are rich, and the complications are traceable to the fact that shortage can be prevented by ultra-violet light, a non-dietetic factor which naturally confused the dietetic issue.

The nutritional road—vitamins (ii) Analysis of diet

The study of deficiency diseases had to be brought into the laboratory from the realm of clinical medicine, where it was less amenable to controlled experiment. The analysis of diet, on the other hand, was from the first a laboratory study. Work during the earlier part of the nineteenth century established protein, carbohydrate, fat, mineral salts and water as the major constituents essential in the diets of higher animals. Before the end of the century, however, voices had been heard questioning whether these alone were sufficient to maintain life and health. Experimental animals such as mice, offered only a basal diet of the purified major nutrients, often failed to survive. The results from different laboratories disagreed, however, for sometimes the animals survived or even grew. We can now guess confidently at the reason; clearly, the purification of the constituents of the basal diet had not been rigorous enough to free them entirely of vitamins—which are, of course, required only in small amounts. Faced with conflicting evidence, the voice of orthodox authority suspended judgment until 1912.

In that year, there appeared in the *Journal of Physiology* a paper which has become famous. It was by Hopkins, of Cambridge, and described his experiments on the growth of rats fed purified, basal diets with and without small supplements of milk. Hopkins used young rats, recognizing that deficiencies were more likely to show during growth. He divided them into two groups, only one of which received the milk supplements. In one experiment, the group getting milk had nearly doubled its average weight after eighteen days, while the other had hardly gained at all. The supplements were now transferred to the other group and, sure enough, the previously failing rats began to put on weight in a healthy way. Those that had been deprived of their milk, on the other hand, soon stopped growing and went into a decline, so that after a further four weeks they had been outstripped in weight by their more fortunate comrades (Fig. 15).

An important feature of Hopkins's experiments was the careful way in which they were controlled. Besides the conviction-carrying device of interchanging his experimental and control groups of animals, he countered a major objection that had been raised against other experiments with purified diets—he checked how much food his rats ate and showed that growth failed even though total consumption was more than enough to cover energy requirements. It was, therefore, not just a question of the animals going on "hunger strike" on the basal diet (which can hardly have been the most appetizing, even by rat standards). Milk must contain hitherto unknown essential food factors, and Hopkins remarked that they must be active in "astonishingly small amounts".

Appearing in the same year as Funk's "vitamine" hypothesis, Hopkins's classical paper first really convinced the world of science that additional food factors do exist. It started the era of intensive vitamin research, and helped in large part to win Hopkins his half-share of the Nobel prize for medicine in 1929 (the other half going to Eijkmann). Yet we now know not only that essentially the same experiments had been done before, but also that they were very difficult to repeat afterwards. In view of their momentous consequences, this adds up to perhaps the sharpest twist of double irony in the history of modern science.

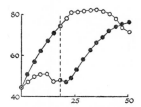

Fig. 15. *Hopkins's experiment to show the presence of vitamins in milk*
The original legend is as follows. "Lower curve (up to 18th day) eight male rats upon pure dietary; upper curve, eight similar rats taking (in addition) 3 c.c. of milk each day. On the 18th day, marked by vertical dotted line, the milk was transferred from one set to the other. Average weight in grms. vertical; time in days horizontal."
The "pure dietary" consisted of purified casein (the major protein of cow's milk), lard, starch, sugar and inorganic salts. The total solids in the milk added to the basal diet amounted to less than 4% of those in the total food.

Results differing little from Hopkins's had been published in 1905 by Pekelharing in a little-known Dutch journal. The experiments were rather less thorough, but showed in a similar way that a purified diet, inadequate by itself for the survival of mice, could be made adequate by adding small quantities of milk. Pekelharing's conclusions were not, however, widely accepted or even known.

Hopkins's experiments are difficult to repeat because, although milk is a good source of some vitamins, it is not particularly rich in others—thiamine, for instance. As little as 2 to 3 ml of milk per day was found adequate in the original work, and this does not supply enough of all the vitamins a rat needs. The reason for the striking success of the 1912 experiments is still not entirely clear. One factor may well be that the basal diet was not entirely free of all the vitamins and contained enough of just those in which milk is poor, so that basal diet and milk complemented each other to make the combination adequate. (If this is the true explanation, it may be added that it was certainly not the last time that biochemical discoveries have depended on the presence of contaminants in supposedly pure substances. Impurities may pay.)

The difficulties about priority and reproducibility detract not one jot, of course, from the correctness and immense importance of Hopkins's general conclusion, out of which grew a large new field of research. By 1915, McCollum and Davis in the United States had shown that at least two factors are necessary—fat-soluble A and water-soluble B. The number of vitamins known continued to increase rapidly during the decades which followed. Proliferation was particularly rapid within the B complex; for historical reasons, virtually all the water-soluble vitamins except C are included in this group. This is highly arbitrary in a way, because the various substances thus lumped together are a most miscellaneous collection, not chemically related to each other at all; but they do tend to some extent to occur together, so that the nomenclature is not entirely without nutritional significance. The B complex includes thiamine, riboflavin, nicotinate, pyridoxine, pantothenate, biotin, folate and B_{12} or cobalamine. Besides the other water-soluble vitamin, called C or ascorbate, there are also the fat-soluble vitamins A, D, E and K.

The enzymological road—coenzymes

The concept of coenzymes arose out of the study of the mode of action of enzymes in extracts of living organisms. Often, it was found, an additional substance has to be present to allow an enzyme to catalyse a change in its substrate. Such substances were called coenzymes, because they work with enzymes to bring reactions about. Unlike enzymes themselves, they are non-protein in nature. Operationally, this means that they can be distinguished from enzymes by being typically thermostable and dialysable. In other words, they are in general not inactivated by boiling, which denatures enzymes (p. 52); and their molecules, though organic, are small enough to pass through the pores of cellophan dialysis tubing (which, very roughly speaking, lets through particles with weights of less than five figures—cf., p. 20).

An early extensive investigation was on "cozymase". The name "zymase", it will be remembered (p. 59), referred to the material responsible for the activity of yeast juice in fermenting sugar to produce alcohol; it is now known to be not a single enzyme but a whole battery of more than a dozen enzymes. Harden and Young, in the course of their classical investigations on the action of yeast juice (which also revealed the surprising phosphate effect—p. 60) found that the power of the juice to induce fermentation is lost if it is dialysed to remove small molecules. Activity can be restored to the dialysed juice, however, by adding either the dialysate (the material that passes through dialysis tubing) or some boiled juice (Table III). Besides the non-dialysable, thermolabile protein fraction, a dialysable, thermostable fraction is thus necesary for fermentation, and it was to this that the name "cozymase" was given.

TABLE III. *Fermenting activity of yeast juice*

Untreated juice	+	Dialysed juice + boiled juice	+
Boiled juice	−	Dialysed juice + dialysate	+
Dialysed juice	−	Boiled juice + dialysate	−
Dialysate	−		

It is now known that, just as "zymase" is a mixture of a number of enzymes, so the activity of "cozymase" is attributable not to one but to several factors.

One of these is NAD, which has already been mentioned (p. 65). Its role is quite well understood—it serves as a hydrogen carrier, accepting hydrogen or giving it up again, as the case may be, in oxido-reduction reactions (such as reactions 6, 12 and 13 of Fig. 11, reactions 6 and 9 of Fig. 14, and the oxidative decarboxylation of pyruvate described on p. 80). From the enzyme's point of view, it is really a second substrate; what the enzyme does is to catalyse transfer of hydrogen (usually in either direction according to the relative concentrations) between it and the other substrate: $AH_2 + NAD \rightleftarrows A + NADH_2$. It is easy to see why the other substrate cannot be acted on in the absence of NAD (or $NADH_2$ as appropriate)—in other words, why the coenzyme is necessary for the enzyme-catalysed reaction to proceed; the enzyme cannot catalyse hydrogen transfer without something to transfer the hydrogen to.

Although NAD can thus in a formal sense be regarded as a substrate, it acts as part of the catalytic equipment of cells. During normal metabolism, it is reduced by a variety of substrates in the presence of their respective dehydrogenases; $NADH_2$ is reoxidized either by other dehydrogenases working "in reverse" (e.g., reactions 12 and 13 of Fig. 11 proceeding in the direction shown) or, in the case of aerobic metabolism, by respiratory chains (p. 78). A molecule of NAD can thus act over and over again, in the manner of a catalyst.

The two roads meet (i) Thiamine, TPP and beri-beri

Another constituent of "cozymase" is thiamine pyrophosphate, called TPP for short. Since this substance illustrates particularly well the relation between vitamins and coenzymes, it is a good case to discuss here in more detail.

TPP is required for the reaction by which pyruvate is decarboxylated to acetaldehyde during alcoholic fermentation by yeast (reaction 11 of Fig. 11).

$$CH_3 \cdot CO \cdot COO^- + H^+ \rightarrow CH_3 \cdot CHO + CO_2$$

The enzyme that catalyses this reaction is called pyruvate decarboxylase. Its action is quite easy to study by adding pyruvate to a yeast extract and measuring either the acetaldehyde or the carbon dioxide produced.

If the extract is dialysed under slightly alkaline conditions, decarboxylase activity disappears because of the loss of a dialysable cofactor. In 1937, Lohmann and Schuster in Germany reported the isolation of this cofactor and the elucidation of its chemical nature, which was relatively easy because it turned out to be nothing more than the pyrophosphate of thiamine. Thiamine itself had already been studied rather thoroughly as the anti-beri-beri vitamin or vitamin B_1; its structure had been determined, and only a few months previously its synthesis in the laboratory had been achieved.

Here, then, the nutritional and enzymological roads met. The vitamin makes up an important—in this case the major—part of the coenzyme. TPP acts by combining with pyruvate; the complex is decarboxylated and then splits to give acetaldehyde and regenerate free TPP which can be used again. The whole process takes place, of course, under the influence of the decarboxylase enzyme.

$$CH_3 \cdot CO \cdot COO^- + TPP \rightarrow [CH_3 \cdot CO \cdot COO^- \cdot TPP]$$

$$\xrightarrow[-CO_2]{+H^+} [CH_3 \cdot CHO \cdot TPP] \rightarrow CH_3 \cdot CHO + TPP$$

In higher animals, the decarboxylation of pyruvate takes place by a more complicated process which in addition involves oxidation (p. 80), but this has been found also to require TPP, as does the closely analogous oxidative decarboxylation of oxoglutarate (step 6 of Fig. 14). (These complex reactions involve several steps and require both NAD and TPP.)

We know a good deal, then, about the action of TPP as a coenzyme. To what extent can we relate this to the action of thiamine as a vitamin? In a general way, it is now obvious why thiamine is necessary for animals (given that they cannot synthesize it themselves). Can we go further in explaining the particular effects of thiamine deficiency on the basis of what is known of the fundamental biochemistry involved? This amounts to a particular case of the general problem of correlating data at the molecular level with data at the whole organism level of organization (p. 97)—an unusually interesting case, for some intriguing suggestions have been made regarding it. They certainly fall far short of a complete explanation, it is true, but they are worth discussing here as steps in the sort of direction that biochemists would like to go.

A good deal of evidence indicates that, in thiamine-deficient animals, the oxidative decarboxylation of pyruvate is impaired. This is a point on the metabolic route-map at which a developing shortage of TPP soon makes itself seriously felt. The bottleneck shows itself by marked rises in the levels of pyruvate (and its reduction product lactate) in the blood and urine; such rises can, indeed, be made the basis of biochemical tests for thiamine deficiency.

The visible symptoms of such deficiency are predominantly nervous. In human beri-beri they include neuritis, weakness and muscular wasting. In the pigeon, a favourite test animal for studies of this type, there is paralysis and the bird characteristically lies with its head sharply bent back ("opisthotonos"). It recovers very quickly, however, when given thiamine; a single injection may make it hold its head up after fifteen minutes and appear normal after an hour. Such "miracle cures" are illustrated in many standard text-books by pairs of photographs of the "before-and-after" type, quite as dramatic as those that used to be popular in the advertising world. Apparently, the primary lesion was what has been called "purely biochemical"—that is, only a lack of a particular kind of molecule. Serious damage at the cellular or a higher level could hardly be repaired so quickly.

How can these macroscopic symptoms be related to the metabolic lesion? All tissues seem to use virtually the same route for oxidizing carbohydrate (though tissues other than muscle can also use to some extent a mechanism alternative to that of Figs. 11 and 14). Why, then, should the nervous system be most affected? One suggestion has been that the clue is its dependence on carbohydrate for its energy supply. Nervous tissue is more limited than others in its ability to use fat or protein, so that it is likely to be hardest hit by a block in carbohydrate breakdown. Other tissues get by better with the oxidation of fatty acids, which does not proceed via pyruvate (Fig. 13). (It does, of course, proceed via the citrate cycle, including the TPP-dependent oxidative decarboxylation of oxoglutarate, step 6 of Fig. 14. However, it has been shown that, with the onset of thiamine deficiency, the oxidative decarboxylation of pyruvate becomes impaired well before that of oxoglutarate. Apparently the system handling oxoglutarate is better at snapping up what it needs from a limited TPP supply.)

Although all this sounds plausible enough as far as it goes, it is clear that interference with the energy supply of the nervous system through impairment of carbohydrate metabolism cannot by itself account for the symptoms of thiamine deficiency. In the case of rats, it has been found that survival is possible for months with almost no thiamine as long as the diet contains little carbohydrate. Only when extra sugar is given does the thiamine deficiency lead to death. Perhaps the abnormal metabolism of carbohydrate gives rise to some substance which itself exerts a harmful effect; indeed, proposals regarding the possible identity of such a toxic agent have been made. Whatever the immediate cause of the deficiency symptoms, nutritionists recommend that the minimum thiamine intake should vary with the amount of carbohydrate eaten; less of the vitamin need be taken if most of the body's energy requirements are satisfied by protein and fat. Beri-beri was endemic among the rice-eating peoples of the Far East. Rice is, of course, a starchy food; when it is highly milled, most of its thiamine is removed with the bran and the rice-eater is in danger of not getting enough of the vitamin to cope with his carbohydrate-rich diet.

The two roads meet (ii) Nicotinate, NAD and pellagra

Thiamine is far from being the only vitamin which, once in the body, is known to enter into the constitution of an enzyme cofactor. Some other cases, all belonging to the B group of vitamins, are summarized in Table IV.

Nicotinate is a case which calls for further comment, if only because the corresponding coenzymes, NAD and NADP, have already come up for mention a number of times. Nicotinate itself has quite a simple structure:—

$$\underset{N}{\bigcirc}\!\!-\!COO^-$$

The name is derived, of course, from the more notorious nicotine, from which the substance too can be prepared, but the similarity in constitution is not quite as close as the names might suggest. Nicotine is considerably more complex, containing a second heterocyclic ring, and vigorous chemical oxidation by nitric acid or potassium permanganate is required to degrade it to nicotinic acid. The mammalian body cannot effect the conversion, and nicotine has no

TABLE IV. *Some vitamins as constituents of enzyme cofactors*

Vitamin	Coenzyme or prosthetic group*	Type of reaction which the cofactor helps to catalyse
Thiamine (B_1)	TPP	Cleavage of carbon-carbon bonds immediately adjacent to carbonyl (i.e., $-\underset{\underset{O}{\|}}{C}-\overset{\|}{\underset{\|}{C}}-$)
Nicotinate	NAD and NADP	Hydrogen transfer.
Riboflavin (B_2)	Flavin mononucleotide (FMN) and flavin adenine dinucleotide (FAD)	Hydrogen transfer.
Pantothenate	Coenzyme A (CoA.SH)	Many reactions involving carboxylic acids; the reactive forms are acyl derivatives of the coenzyme, CoA.S.COR—e.g., acetyl-coenzyme A (p. 80).
Pyridoxine (B_6)	Pyridoxal phosphate	Many reactions of amino-acids
Folate	Tetrahydrofolate	Transfer of the one-carbon fragments —CHO and —CH_2OH
Biotin	?	Carboxylation ($RH + CO_2 \rightarrow RCOOH$)

* The heading "coenzyme or prosthetic group" calls for comment. "Coenzyme" conjures up a picture of a substance distinct from the enzyme which has to be added to enable an enzyme-catalysed reaction to proceed. "Prosthetic group" suggests something different—a conjugated protein, containing in one and the same molecule a moiety of protein, made up of amino-acids, and a moiety made up of something else, the prosthetic group. Such a prosthetic group can participate in a reaction very much like a coenzyme does, with only this exception—that instead of being independent of the enzyme it is permanently attached to its protein partner. The riboflavin derivatives, for instance, which are yellow in colour, are attached to proteins in yellow complexes called flavoproteins (p. 94). Just like nicotinamide in NAD and NADP, the riboflavin portion is the part which reversibly accepts hydrogen in hydrogen transfer reactions. The prosthetic groups FMN and FAD, in other words, are "built-in coenzymes". (So are the haem prosthetic groups of cytochromes (p. 78), though in this case no vitamin forms part of them.)

The distinction between coenzymes and prosthetic groups cannot, however, be made as hard and fast as is implied by this statement of extreme cases. In order to effect a reaction involving a coenzyme, an enzyme molecule must combine with it (as in fact it must with any substrate). The combination may be only momentary, but it must take place, for the enzyme cannot act at a distance—it must get to close grips. In fact, it is found that enzyme-coenzyme pairs have definite and measurable affinities for each other, which can be expressed as the equilibrium constant for the reversible reaction

$$\text{enzyme} + \text{coenzyme} \rightleftarrows \text{complex}.$$

The further this equilibrium lies over to the right, the stronger the binding.

For NAD and the proteins with which it collaborates, binding is relatively loose; hence NAD is commonly thought of as a coenzyme floating around free. For flavoproteins, on the other hand, binding is in general firm; hence FMN and FAD are usually thought of as prosthetic groups of conjugated proteins. But where are we to draw the line between loose and firm binding? It is rather like being asked to decide at what length a piece of string stops being short and starts being long. To pick on any particular numerical value for equilibrium constant or length must be purely arbitrary.

Further complications that arise are illustrated by the case of TPP. To dialyse this away from yeast pyruvate decarboxylase (p. 120), the pH must be slightly alkaline. Below pH 7 it is more firmly bound to its protein partner, and the complex can be isolated as such, containing a fixed amount of TPP per given weight of protein. The animal enzymes which need TPP for the oxidative decarboxylation of pyruvate and α-oxoglutarate bind the cofactor less strongly and are separated from it during purification procedures; to assay their activities, it is necessary to add TPP to the test system.

Criteria other than strength of association have been suggested for distinguishing between coenzymes and prosthetic groups. As long as the issue remains a purely semantic one, concerned with the meanings we choose to give to words, it is of no great moment. It does, however, lead on to a more general point of interest.

Which are to be regarded as the molecular units of the catalytic machinery—enzymes and cofactors separately or in combination? From the above discussion, it emerges that it is not always easy to say which are the individual molecules in the classical sense. Thus a ceiling for biochemistry at the molecular level is not only undesirable but impossible to fix precisely. The borderline between units and complexes is a hazy one.

Extending this line of thought to larger magnitudes and higher levels of organization, the suggestion has been made that subcellular particles in which blocks of a number of enzymes and their cofactors are bound together in apparently specific ways should be regarded as "supramolecules", one stage further up the hierarchy than "macromolecules" such as proteins. The particular context in which this idea has been put forward is that of mitochondrial fragments containing some respiratory chain components (pp. 76, 98). It may well be that the concept of "supramolecules" involves stretching the meanings of words beyond the elastic limit. Nevertheless, it does represent some sort of attempt to get to grips with a problem which biochemistry must eventually tackle—namely, to study not only molecules themselves but also the mutual relations between them in living organisms.

vitamin activity. To prevent any misconception on this point among smokers whose optimism exceeds their knowledge, the name niacin is preferred to nicotinate in the United States of America.

Nicotinic acid occurs in NAD and NADP in the form of its amide, with —$CONH_2$ replacing the carboxyl group. (The abbreviations stand, it will be remembered, for nicotinamide adenine dinucleotide and its phosphate, respectively. The structure of NAD is nicotinamide-ribose-phosphate-phosphate-ribose-adenine. A nucleotide is a structure of the type base-sugar-phosphate (cf. p. 168). In NAD, the bases of the two nucleotide moieties are nicotinamide and adenine respectively, and the sugar is ribose in each case. NADP has a third phosphate residue, which is attached to the ribose residue next to the adenine.)

Nicotinamide is not the largest part of these molecules, as thiamine is of TPP, but it is the functional part, for it is here that the two atoms of hydrogen are added. The nicotinamide moiety is linked to the rest of the molecule (X in the formulae below) through its ring nitrogen, which is quaternary (i.e., tetravalent N^+) as in ammonium salts; one hydrogen is added here, the other at the opposite side of the ring, the *para* position. Since the nitrogen atom is only weakly basic, the base is largely dissociated at physiological pH, losing most of its H^+ into the aqueous medium. To see what these reactions entail, it is best to write the ring formula in full, putting in all the C and H atoms.

$$\begin{array}{c}\text{CH} \\ \text{HC} \diagup \diagdown \text{C—CONH}_2 \\ \| \quad \quad | \\ \text{HC} \quad \text{CH} \\ \diagdown \diagup \\ \text{N} \\ | \\ \text{X}\end{array} \quad \underset{-2H}{\overset{+2H}{\rightleftarrows}} \quad \begin{array}{c}\text{CH}_2 \\ \text{HC} \diagup \diagdown \text{C—CONH}_2 \\ \| \quad \quad \| \\ \text{HC} \quad \text{CH} \\ \diagdown \diagup_+ \\ \text{NH} \\ | \\ \text{X}\end{array} \quad \overset{\text{Spontaneous dissociation}}{\rightleftarrows}$$

$$\begin{array}{c}\text{CH}_2 \\ \text{HC} \diagup \diagdown \text{C—CONH}_2 \\ \| \quad \quad \| \\ \text{HC} \quad \text{CH} \\ \diagdown \diagup \\ \text{N} \\ | \\ \text{X}\end{array} \quad + H^+$$

Nicotinate was identified as a coenzyme constituent before its vitamin function was recognized; the historical sequence of discovery was here the reverse of that in the case of thiamine. Here it was work on coenzymes which helped vitamin research, instead of the other way round.

The discovery of the coenzyme role of nicotinate was due, like much of our knowledge of the factors concerned in biological oxidations, to the work of Warburg (p. 94) in Germany between the two World Wars. (Warburg was awarded the Nobel prize for medicine in 1931.) During studies of an oxidation system in red blood corpuscles, he noted that a thermostable, dialysable cofactor is required and isolated it; it was what is now called NADP. In 1935 he identified nicotinamide as one of the products of its hydrolytic breakdown. A good deal of work had at this time already been devoted to the oxidation-reduction cofactor of yeast "cozymase" (i.e., NAD), but its structure had been only partly elucidated. With the aid of the clue provided by Warburg, however, it did not take long to show the presence of a nicotinamide moiety in this substance too.

Much effort had also been expended by this time on pellagra, which was recognized as a deficiency disease curable by changes in diet, but the nature of the vitamin concerned remained unknown. Here again, Warburg's discovery provided the crucial hint. By 1937 nicotinate was identified as the main "pellagra-preventing factor". (The factors leading to pellagra are, however, rather complicated. In the first place, diets deficient in nicotinate are usually deficient in other vitamins of the B groups as well. Secondly, animals can make nicotinate from the amino-acid tryptophan, a normal constituent of proteins. For the poor sections of the population in the south-east of the United States of America, among whom pellagra was so common, a staple food was maize. Now maize protein is notoriously deficient in tryptophan; the prevalence of the disease was due as much to lack of tryptophan as of nicotinate in the diet.)

We know the coenzymes to which nicotinate gives rise, but it is difficult to pinpoint more exactly the biochemical lesion in pellagra or to relate it to macroscopic symptoms. As nicotinate shortage progressively restricts the supply of NAD and NADP, which of the many reactions known to require one or other of these coenzymes is the first to feel the pinch enough to upset the functioning of the body as a whole? How does this lead to dermatitis, diarrhoea and dementia (p. 115)? Answers to these equestions have been suggested, but they fall short even of the fragmentary thiamine story, both in the detail they offer and in the conviction they carry. Knowledge of the correlation between the molecular and the whole organism levels is almost non-existent here—and the same applies in general to the cases of other vitamins and their corresponding cofactors.

The difficulties of biochemical prediction

The moral of the thiamine and nicotinate stories is a tantalizing one. Interaction between investigations at different levels of organization can be very fruitful; work at one level may throw up decisive clues for another level, whether higher or lower. But contact is difficult to establish at all fully—especially where, as in the case of coenzymes and vitamin deficiencies, correlations have to be made across all the levels intervening between molecules and whole organisms. Only precarious bridges of explanation have been set up between these two extremes. They are not solid enough to give much of the sort of predictive power that science likes to aim at—in this case notably the power to predict, from a knowledge of the fundamental biochemistry involved, the various effects of vitamins or their deficiencies in animals.

Now this is just the kind of predictive power that one might hope for from biochemistry—the power to predict what will happen to an organism if certain substances are administered or withheld. "What will happen to me if I swallow these pills?" is just the sort of question an outsider might reasonably expect a biochemist to be able to answer. Yet even in the relatively well-studied cases of vitamins, far from any appreciable ability to deduce effects in advance, even understanding of the effects after the event is no more than fragmentary. Far from any great *a priori* predictive power, even the *a posteriori* explanations leave much to be desired.

The difficulty is due partly, of course, to gaps that remain in our knowledge of the metabolic reactions in which vitamins take part—that is, to incomplete understanding of events at the molecular level. Even were this understanding more complete, however, knowledge of higher levels of organization would clearly still be necessary to predict the effects observed in whole organisms. In a very real and practical sense, therefore, investigations at all levels of the organization hierarchy are called for. Although a case has been made for considering work at the molecular level to be the most fundamental kind of biological investigation (p. 16), there can be no question of its superseding work on less disrupted systems (pp. 95, 108).

To predict effects of molecules on other molecules is easier, of course, than to predict effects of molecules on living organisms—precisely because no higher levels of organization have to be taken into account. One can, however, afford to make a value judgment here and say that biochemists should rate effects in whole organisms higher in importance than effects in disrupted systems. Disruption may be a necessary part of analytical procedure, but applicability to intact organisms remains the criterion of biological significance. It is, after all, in intact organisms that life is lived, and this is what biochemistry should set out ultimately to explain, rather than laboratory creations of its own. Tactical manoeuvres may profitably be directed at more limited aims, but the grand strategy must be above all to solve the problems posed by nature rather than those set up by biochemists themselves—to explain life, not more or less artificial products of biochemical technique.

Understanding of the effects of substances on organisms—understanding complete enough to give some measure of predictive power—is thus a long-term ideal for biochemical theory to aim at. As yet, obviously, it remains little more than a pipe-dream. The appetite has been whetted by thiamine, but nearly all remains to be done before it can be satisfied. Biochemistry as a predictive science in the sense at issue here is a thing of the far distant future. (Even the ambition is, perhaps, best nurtured in the privacy of biochemical departments, for fear of condescension, ridicule or worse at the hands of other biologists who—for practical reasons in the clinic or in the field, for instance—set themselves more limited objectives and come closer to fulfilling them.)

The ideal is, however, an eminently worthy one, for vitamins are neither the only nor the most urgent case in point. Another, more pressing challenge of the same type arises in the case of drugs and their effects on host as well as parasite. It seems appropriate, therefore, to take up this analogous problem as the topic of the next chapter.

VII

Designing Drugs—The Dream of a Rational Chemotherapy

Old ideas behind old remedies

IT has been said that, whereas chemotherapy as an art is as old as civilization, chemotherapy as a science is the child of to-day. If this is so, we may still be left wondering whether the birth-pangs are yet over. When one speaks of a science, after all, one thinks of a body of knowledge with some degree of rational coherence which gives it a certain amount of predictive power. How far has chemotherapy progressed towards this goal? This is the main question discussed in the present chapter.

It seems hardly necessary to emphasize how difficult an undertaking it is to reach a thorough understanding of drug action. Just as in the case of vitamins and the symptoms of the deficiency diseases they prevent, we are concerned with the effect of chemical substances on whole organisms. Once again, therefore, we are confronted with the yawning chasm of ignorance which separates our knowledge of the molecular level of organization from that of the intact animal. The organization of a living organism is highly complex. Introducing a strange kind of molecule may well alter its basic biochemistry in more than one way; even in the unlikely event that only one metabolic reaction is markedly affected, this can show itself as a bewildering array of the sort of symptoms with which clinical medicine has to concern itself.

Even the most basic fact about a drug is by no means simple to establish. "Does it work?" is a question that is easier to ask than to answer. We may well be tempted to smile at some of the curious remedies that found favour in former times. But physicians must inspire confidence in their patients; so it is not merely a matter of intellectual dishonesty that the medical profession has always cultivated the impression that cures are because of, but deaths despite, the treatment given. To get at the real cause-and-effect relationships requires more than just a little straight thinking. Modern pharmaceutical firms employ highly trained scientists and lavish on them all the resources that money can buy without achieving infallibility. The second half of the twentieth century has already seen its share of drugs hailed as miracle cures one year, and condemned the next as useless, or dangerous, or both.

A glance at an old pharmacopoeia, therefore, astonishes not so much by the extent of human credulity that it reveals as by the witness it bears to the fertility of human invention. "Usnea" was an official drug in the pharmacopoeia until the nineteenth century, and credited with impressive powers of healing; it was made from the moss scraped from the skull of a criminal who had hung in chains. Other sources of drugs in relatively recent European use include crocodile dung, Egyptian mummy and even unicorn's horn.

Ever since men began to think systematically about natural phenomena, they have tried to see their way through the maze of drugs and diseases by the light of some kind of rational theory. The form that their theories took naturally depended on the views they held as to the composition of the body—the prevailing opinions about biochemistry, one might say, anachronistically transplanting the term backwards through the centuries. In ancient Greece, for instance, it was usual to think of all substances—living and non-living alike—in terms of degrees of the qualities hot or cold, and wet or dry. (In this way, the Greeks arrived at their famous four "elements", earth, water, air and fire—earth being cold and dry, water cold and wet, air hot and wet, fire hot and dry. These "elements" obviously stood for a notion very different from that of a modern chemical element. They were not nearly as useful, not remaining constant in quantity, as a modern element does under all but very abnormal circumstances; but the idea was reasonable enough as far as it went, since obviously all substances can be placed on scales of heat or cold and of moisture or dryness—in the same way that any point on a plane can be described in terms of two quantities, giving the ordinate and the abscissa respectively). Health was attributed by the Greeks to a satisfactory and harmonious balance or "temperament" of the four primary qualities, and disease to a disturbance or imbalance in their proportions. Remedies, accordingly, were designed to redress the balance. Each of the great variety of herbs which made up the bulk of the ancient *materia medica* was assigned degrees of heat and moisture (on the basis of criteria which often seem far from obvious to modern readers), and "cold" drugs were given to alleviate "hot" diseases; for what, it was argued, could be more obvious than that a symptom must be cured by its opposite? This whole system of therapeutics was given full expression by Galen, the great medical writer of the second century A.D., whose authority was not seriously questioned until the Renaissance and thus formed the basis of learned European medical practice for more than thirteen centuries.

In the seventeenth century, an era of vigorous and often fruitful scientific thought and endeavour, a different theory became fairly popular. According to the new view, diseases are caused by imbalance between acid and alkali in the body fluids. Therapeutics now became a matter of counteracting opposite with opposite according to the new duality—either by administering acid or alkali as such, or by feeding diets supposed to predispose to acidity or alkalinity. Smallpox, for instance, was thought of as a disease of alkalinity, and the greatest medical authorities of the eighteenth century continued to extol the curative virtues of sulphuric acid in the drink.

To-day, we are confident that our knowledge of bodily constitution and function is more soundly based and more detailed than ever before. Can we, as a result, propose a better rationale for therapeutics?

The importance of selectivity

Modern chemotherapy owes much to the work of Ehrlich in Germany during the early years of the present century. The arsenicals that developed out of his efforts have largely been superseded by newer and better drugs, but his work remains instructive as an attempt to put pharmaceutical research on a more rational basis.

Ehrlich has been called the "father of chemotherapy", and he can certainly claim paternity at least as to the name, which he introduced. The aim of chemotherapy is to use drugs to attack parasites within the host. (The word chemotherapy is often used as Ehrlich meant it, to refer to the treatment of infectious diseases only, although in itself it implies no restriction to exclude non-infectious diseases). Ehrlich pointed out that the fundamental concept of chemotherapy must be that of *selective toxicity*. The emphasis here is on the *selectivity*. It is easy enough to kill micro-organisms; the difficulty is to kill them without harming the host. In more precise terms, a successful drug must have a high chemotherapeutic index—the ratio of the maximum dose tolerated by the host to the minimum dose effective against the pathogen.

$$\text{Chemotherapeutic index} = \frac{\text{maximum tolerated dose}}{\text{minimum effective dose}}$$

Clearly, the chemotherapeutic index is a measure of the margin of safety in the use of a drug. It indicates by how much one can afford to exceed the minimum curative dose without harming the patient.

To appreciate the direction which Ehrlich's efforts took, it must be remembered that, by the beginning of this century, the germ theory of disease had borne its first major crop of fruit in the form of immunological means of combating infections. Inoculations and antisera were already capable of coping with a number of bacterial diseases, and many experts believed—over-optimistically, as it turned out—that the rest would soon succumb. It was also realized, however, that many disease-producing micro-organisms other than bacteria fail to elicit a good response of antibodies. The need for new drugs seemed most pressing in the fight against such non-bacterial pathogens. Most of Ehrlich's work was, in fact, directed against trypanosomes, protozoa which cause serious tropical diseases of cattle and sleeping sickness in man; and against spirochaetes, spirally-shaped organisms like bacteria but not classified as true bacteria, which include *Treponema pallidum*, the organism that causes syphilis.

Earlier in his career, Ehrlich had done immunological work (in fact, his half-share of the Nobel prize for medicine in 1908 was principally for his contributions in that field). He was much impressed by the way that, in infected animals, antibodies seemed unerringly to single out the parasites while leaving host tissues unscathed. An antibody is a special protein in the blood serum (cf., Fig. 9) which is formed by an animal in response to an antigen, a foreign substance which may derive, for instance, from a bacterium. A given antibody can combine with the antigen that evoked its formation, thereby neutralizing it, but it does not combine with other substances (unless they are very closely related to the antigen). Precisely this property of selectivity was what Ehrlich wanted to copy in his synthetic drugs. He regarded antibodies as the best therapeutic agents—"magic bullets", as he called them, which find their targets by themselves without hitting the innocent bystanders, the host cells. It was where this ideal of "serum therapy" was not practicable that he wanted chemotherapy to emulate its accuracy. His objective was, in effect, to make artificial "bullets", aimed by chemical means—by being given a structure that selectively attacks the disease-producing micro-organisms.

From the mice that "died cured" to organic arsenicals

It was already known that arsenic compounds show the right kind of activity. Trypanosome infection in mice could be eradicated by arsenious acid. The snag was that the mice died too; but at least they "died cured" of the infection, which amounted to the hint of a promise.

In 1905, Thomas and Breinl at Liverpool found a substance that could cure without killing. Instead of a simple, inorganic arsenic compound, they used an organic derivative containing an aromatic ring linked to an arsonic acid group, —AsO_3H_2. To indicate its relative lack of toxicity, it was called "atoxyl" (Fig. 16), but the

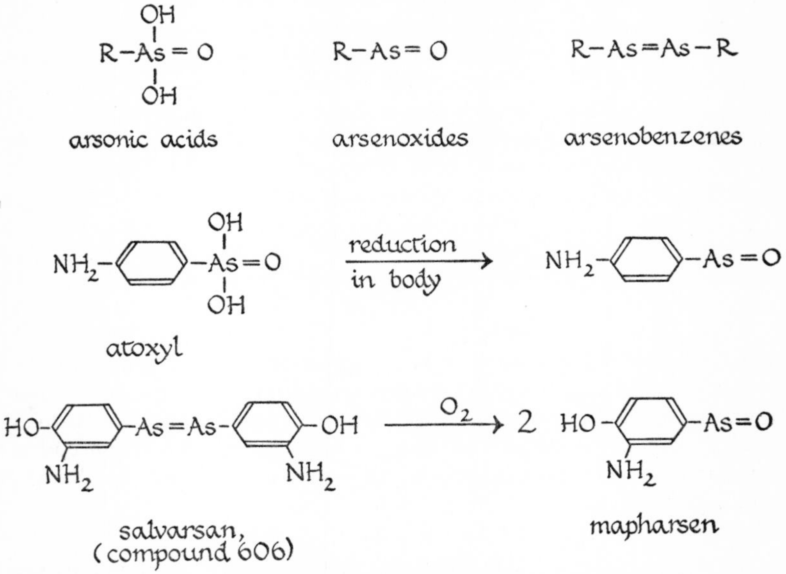

$$R-\overset{\overset{OH}{|}}{\underset{\underset{OH}{|}}{As}}=O \qquad R-As=O \qquad R-As=As-R$$

arsonic acids arsenoxides arsenobenzenes

$$NH_2-\langle\bigcirc\rangle-\overset{\overset{OH}{|}}{\underset{\underset{OH}{|}}{As}}=O \xrightarrow[\text{in body}]{\text{reduction}} NH_2-\langle\bigcirc\rangle-As=O$$

atoxyl

$$HO-\underset{NH_2}{\langle\bigcirc\rangle}-As=As-\underset{NH_2}{\langle\bigcirc\rangle}-OH \xrightarrow{O_2} 2\ HO-\underset{NH_2}{\langle\bigcirc\rangle}-As=O$$

salvarsan, mapharsen
(compound 606)

Fig.16. *The three types of arsenical drugs. Both arsonic acids and arsenobenzenes are convertible into arsenoxides, which are the active forms*

name seems over-optimistic in retrospect; the drug has a low chemotherapeutic index and its use is liable to cause blindness.

Following up this lead, Ehrlich set his team of workers at his Institute in Frankfurt to investigate many other organic arsenic compounds. An indication of the scale of their work is the serial number 606 of the compound which achieved fame; although this may seem puny compared to the testing programmes of modern pharmaceutical firms, it was impressive by the standards of 1909, the year that saw the first encouraging results with salvarsan. This was an arsenobenzene derivative, in which two arsenic atoms linked to each other each carry an aromatic ring (Fig. 16). Both trypanosome and spirochaete infections responded to it, in humans as well as experimental animals. Compared to earlier drugs, it was of a new order of effectiveness, and mankind's pent-up emotions about the scourge of syphilis were released in a flood of gratitude over Ehrlich.

One of the principles which guided Ehrlich in his work was the belief that good drugs act directly on the parasites. It was a belief which did not go unchallenged, for many preferred to attribute cures to a strengthening of the hosts' defensive mechanisms; and for a time there was some basis for this view of the mode of action of the arsenicals, for little direct action on the parasites could be shown—trypanosomes and spirochaetes in the test tube seemed relatively indifferent to quite high concentrations of both atoxyl and salvarsan. Ehrlich was able to show, however, that trivalent arsenoxides, substances of the general formula R—As=O, are active *in vitro*. He suggested that the arsonic acid group of atoxyl is reduced in the mammalian body to give such an arsenoxide.

$$\begin{array}{c} \text{OH} \\ | \\ \text{R}-\text{As}=\text{O} \\ | \\ \text{OH} \end{array} \xrightarrow[\text{in body}]{\text{reduction}} \text{R}-\text{As}=\text{O}$$

During the nineteen-twenties Voegtlin, of the United States Public Health Service, confirmed that such reduction of arsonic acids does take place. Furthermore, he showed that arsenobenzene derivatives like salvarsan undergo oxidation, likewise giving arsenoxides. Similar oxidation takes place readily even on shaking an alkaline solution in air.

$$\text{R}-\text{As}=\text{As}-\text{R} \xrightarrow{O_2} 2\,\text{R}-\text{As}=\text{O}$$

The importance of this work lay in the attention it drew to the possibility that a drug may not be effective in the form given but be converted first into another compound which is the active one. Of the three groups of arsenical compounds concerned—the arsonic acids with their pentavalent arsenic, the arsenoxides with their trivalent arsenic, and the arsenobenzenes with trivalent arsenic further reduced—the arsenoxides are the ones which actually exert the effect against the parasites.

Their mode of action seems to be on sulphydryl groups, with which they react reversibly:—

$$\text{R}-\text{As}=\text{O} \quad \begin{array}{c} \text{HS}- \\ \\ \text{HS}- \end{array} \rightleftarrows \text{R}-\text{As} \begin{array}{c} \diagup \text{S}- \\ \\ \diagdown \text{S}- \end{array} + \text{H}_2\text{O}$$

Sulphydryl groups are essential in many enzyme reactions. Sometimes they are present in the enzyme molecules themselves, where they occur in the side-chains of cysteine residues (Table I); in other cases, cofactors for enzyme reactions carry vital —SH groups. In a general way, therefore, it is quite easy to see that interference with —SH groups may seriously affect metabolism. Also, it is understandable that compounds containing —SH groups temporarily neutralize the effects of arsenoxides. Thus, Voegtlin showed that if mapharsen, the arsenoxide derived from salvarsan (Fig. 16) is given to a rat infected with trypanosomes, the number of pathogens in the blood immediately begins to decline drastically; if an —SH compound is given with the mapharsen, however, the decline is delayed for several hours. The effect of the —SH compound is attributable to its mopping up the arsenoxide, temporarily keeping it away from the —SH groups vital to enzyme systems.

Host organisms, of course, also have —SH groups which are essential for their metabolism, and it is not surprising that all arsenicals, including salvarsan, are to some extent toxic. The secret of the selectivity of successful drugs, which makes them more poisonous to trypanosomes or spirochaetes than to mammals, lies in the aromatic ring and its substituents; inorganic and aliphatic arsenicals, lacking such a ring, are quite as toxic to host as to parasite. Inorganic arsenic compounds, of course, are notorious for the popularity they have enjoyed through the centuries for both homicide and suicide. An example of an aliphatic arsenical is the war-gas Lewisite. This, too, is not selectively toxic; it is a poison gas, not a drug, and it kills friend and foe equally, whether microbial or human. The unpleasant possibility of its use in war has stimulated interest in possible antidoes to it, and —SH compounds naturally suggested themselves as candidates to fulfil this role. Simple —SH compounds proved not to be effective enough, but work at Oxford during the Second World War showed that compounds containing two —SH groups in close vicinity are good antidotes. The substance called BAL (an abbreviation for British Anti-Lewisite) is of this type; it is good at mopping up the poison because the combination results in ring formation:—

$$Cl\cdot CH=CH-As{\begin{matrix}Cl\\Cl\end{matrix}} + {\begin{matrix}HS-CH_2\\HS-CH\\HO-CH_2\end{matrix}} \rightarrow Cl\cdot CH=CH-As{\begin{matrix}S-CH_2\\S-CH\\HO-CH_2\end{matrix}} + 2\,HCl$$

Lewisite   BAL

From dyes to sulphonamides

Arsenicals, it will be remembered, were developed for use against microbes other than true bacteria. It was not until the nineteen-thirties that the first good antibacterial drugs, the sulphonamides, became available.

The crucial lead from which their discovery followed came from the dyestuff industry. This industry has, of course, long been a major source of new organic compounds. Besides arsenicals, some dyes and related products had been found effective against trypanosomes, and firms producing dyestuffs were in the habit of routinely testing their new products for chemotherapeutic activity. In the early nineteen-thirties Domagk, of the German firm I. G. Farbenindustrie, showed that a red dye called prontosil is effective against streptococcal infection in mice. His observation became so important that he was awarded the Nobel prize for medicine in 1939. (He had to decline it, though, for Germans were forbidden to accept by a decree issued by Hitler in 1937.)

The formula of prontosil is fairly complex; it is worth looking particularly at the right-hand half of the molecule as written below:—

$$NH_2-\langle\bigcirc\rangle-N=N-\langle\bigcirc\rangle-SO_2NH_2$$
$$NH_2$$

Streptococci growing in the laboratory outside a host were found not be affected by prontosil. A similar situation had arisen with arsenicals, of course, and the lessons of the earlier work were now turned to good use, suggesting as they did that the prontosil molecule is changed in the mammal to give rise to the active substance. It was soon shown, at the Pasteur Institute in Paris, that prontosil is in fact split by reduction in the body into two halves; the half active against bacteria is sulphanilamide:—

$$NH_2-\langle\bigcirc\rangle-SO_2NH_2$$

This was a vital observation. It directed future research to making more sulphanilamide derivatives rather than more azo dyes. Dyestuff chemists might not have expected the activity to reside in this part of the molecule; considered from the dyeing point of view, after all, the sulphanilamide portion of prontosil is merely an accessory—it does not confer colour, but merely acts as a polar anchor to help it bind to wool. Moreover, I. G. Farbenindustrie had patent protection for prontosil, but that did not prevent others from making derivatives of sulphanilamide. Many such derivatives were soon prepared—by 1945, more than five thousand had been described. These sulphanilamide derivatives are the sulphonamides. Most of the effective ones have the general formula

$$NH_2-\!\!\left\langle\!\!\bigcirc\!\!\right\rangle\!\!-SO_2NHR$$

where R is one of a large variety of heterocyclic rings. The first really good one to be produced was sulphapyridine, where R is pyridine; sulphadiazine is one in wide use now.

The sulphonamides were the mainstay of antibacterial chemotherapy during the Second World War. They have now been ousted to some extent by antibiotics, but remain important, being cheap and easy to administer.

The importance of competition—an imperfect key to block the keyhole

The discovery of the sulphonamides, it can be seen, was pretty much an empirical, hit-or-miss affair. One out of thousands of dyes, tested more or less at random, turned out to be active. Soon, however, indications appeared as to the mode of action of the sulphonamides, and a little of the light of reason was shed upon the scene.

The crucial discovery, made by Woods in London in 1940, was that the antibacterial effect of sulphonamides is reversed by the substance para-aminobenzoate (PAB for short). This effect is competitive—sulphonamide stops the bacteria from growing, PAB makes them grow again. The outcome, in terms of bacterial growth, depends less on the absolute concentrations of either substance than on the ratio between them.

To illustrate this important effect, it is best to take some actual figures obtained experimentally (Table V). It is necessary to be quite clear to what these figures refer. The left-hand column gives the molar concentration of sulphanilamide added to the bacterial growth medium. In the absence of any PAB, any of these concentrations are enough to stop the streptococci from growing. If enough PAB is also added, growth can start again. The minimum concentration of PAB necessary to enable this to happen is given in the middle column; it is found always to stand in a fixed ratio to the sulphanilamide concentration, even though the latter varies over a wide range. In other words, one part of PAB counteracts 5,000 of sulphanilamide whatever the absolute concentrations. With the more potent sulphanilamide derivatives, the sulphonamide drugs which have since come into use, similar results can be obtained, except that they are more powerful growth inhibitors, and more PAB is required effectively to antagonize their inhibitory action.

Now competition effects of just this kind are well known in enzymology. It is quite common for enzymes to be competitively inhibited, and the competitive inhibitor is always structurally similar

TABLE V. *Competition between sulphanilamide and PAB*

Sulphanilamide concentration	PAB concentration required to reverse inhibition of growth	Ratio sulphanilamide/PAB
0.3×10^{-3} M	0.6×10^{-7} M	5,000
1.5×10^{-3} M	3×10^{-7} M	5,000
7.5×10^{-3} M	15×10^{-7} M	5,000

Based on the data of Woods, 1940; the numerical values have been rounded off.

to the substrate. For instance, the dehydrogenation of succinate, catalysed by the enzyme succinate dehydrogenase (Fig. 14), is competitively inhibited by malonate.

$$\begin{array}{cc} COO^- & COO^- \\ | & | \\ CH_2 & CH_2 \\ | & | \\ CH_2 & COO^- \\ | & \text{Malonate} \\ COO^- & \\ \text{Succinate} & \end{array}$$

The specificity of succinate dehydrogenase for succinate—its refusal to act on other compounds—must be due to the fact that a part of its molecule, its "active centre", is adapted to combine with substances of the type of succinate. Enzyme and substrate structures must in some way be complementary, like lock and key. Malonate resembles succinate sufficiently to combine with the active centre, but it cannot be acted on. By displacing succinate from the active centre, it merely impedes the dehydrogenation. In terms of the lock-and-key analogy, it is an imperfect key which fits the lock only well enough to block the keyhole. The effect is competitive because malonate and succinate jostle each other for the limited number of places on active centres, in a sort of molecular version of the game of musical chairs.

Sulphonamides resemble PAB almost as closely as malonate resembles succinate.

$$NH_2\text{-}C_6H_4\text{-}SO_2NHR \qquad NH_2\text{-}C_6H_4\text{-}COO^-$$
$$\text{Sulphonamide} \qquad\qquad\qquad \text{PAB}$$

It does not seem too wild a guess, therefore, that sulphonamides act by displacing PAB from an important enzyme.

No metabolic role for PAB was known in 1940, but Woods predicted that one would be found, and his prediction was right. PAB forms part of the molecule of folate which, in its reduced form, tetrahydrofolate, is an important enzyme cofactor (Table IV). (The structure of folate is best looked on in three parts. First, there is a complex heterocyclic ring system of the pterin type, related to some butterfly wing pigments; this is linked to a PAB residue, and this in turn to a residue of glutamate (Table I)). Apparently, sulphonamides act by interfering with the build-up of folate from PAB.

$$\text{PAB} \xrightarrow{\text{sulphonamide} \;\;|} \text{folate}$$

The lock which they block seems to be an enzyme concerned with the utilization of PAB for this important synthesis.

Animal cells, of course, also require folate for some of the reactions vital to their metabolism. Their insensitivity to sulphonamides seems to be due to the fact that they use preformed folate rather than making it themselves from PAB, remaining thereby independent of the sulphonamide-sensitive synthesis; this difference is, apparently, the basis of the selectivity of sulphonamide action. Similarly, those bacteria which require folate to be supplied to them ready-made, and will not grow in a folate-free medium, are mostly unaffected by sulphonamides; it is, in general, only those which use PAB that respond to these drugs.

A rationale at last? The antimetabolite rush

With the sulphonamides, then, we now have a reasonably satisfactory rational basis for the chemotherapeutic action. Some of the detail is still missing, it is true, but the outline seems sound as far as it goes. Does this understanding of the action of an important class of drugs open up vistas of a whole new rational chemotherapy?

In the nineteen-forties, it seemed that it might. The idea was that new drugs should be sought among other "antimetabolites", chemical analogues of normal metabolites that would interfere with their functioning in the sort of way that sulphonamides interfere with the functioning of PAB—more slightly imperfect keys to block other metabolic locks, that is. It was a fine, tempting hare of an idea, and soon the laboratory hounds were in full cry after it. There was no difficulty in preparing antimetabolites—they were soon known in their hundreds. Slight chemical alterations in vitamin molecules often produce substances with antagonistic effects; for instance, 3-acetylpyridine is an antagonist of nicotinate.

nicotinate   3-acetylpyridine

Antagonists were developed not only for vitamins, but also for aminoacids and other items of the chemical equipment of cells.

From the chemotherapeutic standpoint, however, the results of this "antimetabolite rush" were disappointing. Most antimetabolites lack selectivity—they do not discriminate in their toxic effects between host and parasite. One is up here against the fundamental biochemical unity that underlies all forms of life (cf. p. 66). Biochemistry so far has been best at revealing the broad features of metabolism, which are common to organisms of the most varied kinds. Most of the metabolites on which attention has been concentrated are part of the most basic biochemical mechanisms, on which cells of all kinds rely. The requirement for selectivity, however, demands that a rational chemotherapy must base itself on biochemical *differences*. To find the differences of detail requires more minute investigation, and remains largely a matter for the future. From the standpoint of basic biochemistry, it is still quite difficult to distinguish between a mammal and a micro-organism. (This is one of the ironical things about biochemistry, as compared to most other kinds of life science—ironical because to the classical biologist, as to the man in the street, it is precisely its infinite diversity that is apt to be the most striking thing about life.)

One useful drug did emerge, in a semi-rational way, from the search for drugs among antimetabolites. Its discovery arose out of the observation that the causative organism of tuberculosis, *Mycobacterium tuberculosis*, is peculiar in being able to use salicylate as an oxidizable substrate. Lehmann, in Sweden, examined a number of analogues of this compound and in 1946 found the para-amino derivative to be effective against tuberculosis.

$$\text{para-aminosalicylate (PAS)}$$

(structure: benzene ring with COO⁻, OH, and NH₂ substituents)

However, salicylate does not antagonize the action of PAS, as it should do if there were a competition here as there is between PAB and sulphonamides. The "rational" route to PAS, therefore, was not really valid—though fortunately this does not decrease the value of PAS as a drug.

Shifting the burden of design—antibiotics

Most of the recent drugs successful against infectious diseases are antibiotics. An antibiotic is the product of one micro-organism active against another; in looking for drugs among antibiotics, therefore, the burden of designing active molecules is shifted from man to micro-organism. The discovery of useful antibiotics has been a matter of largely unrelieved empiricism—it has been made systematic, but hardly rational. Man's role in their development for clinical use has been to pick and choose, to purify and on occasion to modify.

Penicillin was the first really successful antibiotic. The story of its discovery is well known. Fleming, in London in 1928, noted that bacterial growth was inhibited on a culture plate accidentally contaminated by a mould; the isolation of the active principle produced by the mould was taken up at the beginning of the Second World War by Florey and Chain at Oxford, to earn them, together with Fleming, the Nobel prize for medicine in 1945.

Other antibiotics have been found by systematically "screening" large numbers of different micro-organisms for antibacterial activity. Even among antibiotics, of course, suitable selectivity between host and parasite is comparatively rare. Hundreds of antibiotics are known, but few are sufficiently non-toxic to mammals. In particular, the antibiotics produced by bacteria (exerting their effects against other kinds of bacteria) are too poisonous for most clinical uses. Successful antibiotics have come mostly from moulds, especially various species of *Streptomyces*. From one such species, *Streptomyces griseus*, Waksman in the United States isolated streptomycin in 1944; he received the Nobel prize for medicine in 1952. From other species of *Streptomyces*, found originally in soil samples, have come valuable drugs of the tetracycline group—aureomycin (chlortetracycline) and terramycin (oxytetracycline).

The role of human chemical ingenuity in this field has gone little beyond tinkering with the structures designed and produced by nature and found by trial and error to have the right kinds of activity. Such tinkering is done in the hope of improving various pharmacological characteristics, notably of eliminating undesirable side effects. The penicillin nucleus (a curious structure of two fused heterocyclic rings, one four-membered and one five-membered) can be equipped with different side chains to make different penicillins. Streptomycin can be reduced to dihydrostreptomycin, which is better than the original in some ways, though worse in others. The chlorine atom can be knocked out of aureomycin to give achromycin (tetracycline), which is somewhat safer to use. This whole approach is less a matter of producing new drugs than of modifying old ones—like making a high performance version of a production car, or changing its body from saloon to estate car type.

Although means of combating most of the serious bacterial infections are now available, it is important to realize that the search for new drugs cannot yet be given up. If a time ever comes when new drugs are no longer needed, it is certainly still far off. The reason lies in the phenomenon of acquired drug resistance. Many bacteria are wily enough to become resistant to drugs to which at first they responded. To control them then, other drugs are required, so that the physician needs ample reserves for his pharmaceutical armoury; and in so far as continued use of drugs is bound to multiply cases of acquired resistance, it is necessary to run quite hard in order merely to stand still in the fight against pathogenic micro-organisms.

We have reason to be grateful, then, for the many man-hours that have been and continue to be spent on the search for new drugs, however empirical it may remain. If research had waited for rational theories on which to base itself, mankind would still lack much of the protection against infectious diseases that it enjoys at present.

VIII

A Common Currency for Energy Transactions—ATP

The convenience of a common currency

PRIMITIVE societies run on a barter basis, but economic development cannot proceed far without the convenience of money. The virtue of a monetary system is that it offers a common medium of exchange. Individuals can be rewarded for work of many different types in the form of a single type of article, with which in turn they can purchase a great variety of goods and services.

Metabolism, in its energy aspect, makes use of a similar simplifying device. To a large extent (though not exclusively) energy is channelled through a single type of molecule—adenosine triphosphate or ATP (p. 64), which acts as the common currency for energy transactions of various kinds.

By this means, considerable flexibility in the arrangements for energy supply is made possible. As is well known, the energy requirements of a mammal can be met by oxidation of any of the three major classes of foodstuffs, proteins, carbohydrates or fats (as well as by some relatively unorthodox dietary components such as alcohol); any or all of these can provide the energy for the processes for which an animal requires energy—for the chemical synthesis involved in repair and growth, for pumping materials from one place to another, and for locomotion. Considered from the standpoint of biochemical engineering, this is no mean feat of design; the animal might be compared to a machine that can act as chemical factory, pump and automobile while burning coal, gas or oil indifferently.

Such versatility could hardly be achieved except by interposing a single mediator between many providers and many users of power. The oxidation of foodstuffs does not directly drive the various energy-requiring processes, but is conducted in such a way as to lead to the synthesis of ATP, which then acts as the immediate energy source for various kinds of work.

THE BIOCHEMICAL APPROACH TO LIFE

Where the value lies—high energy bonds

What enables ATP to function in this useful way? The answer lies in the energy which can loosely be thought of as associated with the bonds between its phosphate residues. This energy, which amounts to some 11 to 12 kilocalories per mole, is often symbolized by writing the bonds as "squiggles":—

adenosine—Ⓟ∼Ⓟ∼Ⓟ

Like other forms of symbolism, this one becomes misleading if taken too literally—the energy is not physically located in the bonds. What the symbolism does express very neatly is the fact that, when the bond is broken, 11 to 12 kilocalories of energy are liberated

adenosine—Ⓟ∼Ⓟ∼Ⓟ $\xrightarrow{\text{hydrolysis}}$ adenosine—Ⓟ∼Ⓟ + HOⓅ + 11 to 12 kcal

or ATP $\xrightarrow{\text{hydrolysis}}$ ADP + HOⓅ + 11 to 12 kcal

Two qualifications have to be added to this description. One concerns the way of breaking the bond, the other the way of reckoning the energy.

Breaking, strictly, might mean putting the molecule in a vacuum and pulling the pieces a long way apart. The energy change involved in such a process might be considered the bond energy in a very fundamental sense—but not in a very practical one, because nothing that occurs in living organisms even remotely resembles such a way of breaking the molecule. ATP is split by *hydrolysis*, and in a biochemical context it is the energy change on *hydrolytic* cleavage that is referred to as a "bond energy".

As for energy, the word stands, of course, for only one basic notion, but there are different ways of calculating the energy change for a process. In the same way—to go back to the financial analogy—there are different ways of calculating profit or loss on a business deal, and skilled accountants can often, without dishonesty, show surplus or deficit at will from the same set of figures. Is one to allow—and if so, how much—for various inevitable concomitants of the deal, such as tax, overheads and interest on capital? For different purposes, different ways of looking at the figures may be useful. Similarly, the *total* energy change for a process is not, in general, the most valuable thing to know about it. For many purposes, it helps much more to know the change in what is called the "free energy" (more specifically, the Gibbs free energy, after the nineteenth-century American scientist Willard Gibbs). This is the energy change suitably adjusted for certain factors unavoidably accompanying the process (the work done against or by the atmosphere due to a change in volume, and the change in entropy, a measure of the randomness of a system).

What makes the free energy change such a good one to focus attention on is that it gives the maximum useful work that a reaction can do. The figure of 11 to 12 kilocalories for the hydrolysis of ATP to ADP is a value for the free energy change, not the total energy change. It shows that one mole of ATP, in breaking down to ADP, can contribute up to 11 to 12 kilocalories for some kind of work in the organism. (It is worth explaining why the numerical value is left approximate. Free energy changes depend on the concentrations of the reactants; since the concentrations of ATP, ADP and inorganic phosphate *in vivo* vary according to the state of activity, no precise figure can be given. The value quoted is for concentrations approximating to the physiological, which are considerably lower than the 1·0 M taken as "standard" by physical chemists.)

The terminal phosphate bond of ATP is often called a high energy bond, or an energy-rich bond, because some other types of phosphate esters on hydrolysis liberate smaller quantities of energy of around 3 to 5 kilocalories. Biochemists talk a great deal about such high energy bonds, and the concept is, indeed, a very important one for understanding the energy aspect of metabolism. Here again, however, it is important not to be misled by putting too naïve a construction on the words. ATP is not a wonder-molecule supercharged with energy by the unique powers of life. All the talk of high energy bonds sometimes gives the impression that ATP might explode when struck, or glow in the dark, or give some other sign of being packed with more energy than it is able comfortably to hold. This is not quite the case. ATP can be bought from any supplier of biochemicals, usually in the form of its sodium or barium salt; it is sent, without any special precautions, as a white powder that lies in its little bottle as placidly as if it were sodium chloride. In aqueous solution, it is true, it is not too stable, but plenty of less stable compounds are known.

Two points have to be borne in mind in this connection. Firstly, hydrolysis energies of 11 to 12 kilocalories per mole do not, in themselves, seem particularly remarkable to physical chemists. What gives ATP such special interest in biochemical eyes is not that it has an extraordinarily large amount of energy, but that it supplies energy in a form that can be used directly in biological processes.

In other words, the crucial point about the energy in the terminal phosphate bond is not that it is enormous, but that it is immediately available for the sorts of things that living organisms have to do.

Secondly, there is a very fundamental point about the sort of information given by thermodynamic data like free energy changes. They refer to initial and final states, but not to routes of getting from one to the other; although they indicate, unequivocally and finally, whether it is feasible, from the energy standpoint, for a certain

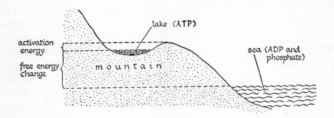

Fig. 17.

The fact that it is thermodynamically possible for the lake water to run down does not mean that it will necessarily do so.

process to take place, they do not indicate whether or how fast it will actually take place. In other words, they show what *can* happen, not what *will*.

The point can be made clear by comparison with a mountain lake (Fig. 17). The height of the lake above sea level corresponds to a free energy change. Knowledge of it shows that, given the chance, water can flow from lake to sea, but not the other way round, unless work is done to lift it up. If the lake water is allowed to run down to the sea, it can on the way be made to do useful work, such as driving water wheels or electric generators. As long as the lake is surrounded on all sides by impenetrable mountains, however, its water never does run down. Its height measures a tendency to run down which may never be realized in practice, unless a suitable route is opened. ATP corresponds to the lake, the splitting of a phosphate bond to the running down of the water. The splitting of ATP can be made to do work; conversely, phosphate cannot combine with ADP unless work is done on the system. But ATP is contained in its little nest of stability by energy barriers which have to be surmounted before breakdown can actually take place. Catalysis by an enzyme reduces these barriers sufficiently for reaction to occur. (The barriers are the

A COMMON CURRENCY FOR ENERGY TRANSACTIONS

activation energy which a molecule has to possess before it can actually react; the catalyst reduces this to a lower level, so that a higher proportion of molecules acquires enough energy to surmount the barriers.)

Pennies for the slot machines—exchanging glucose for ATP

Glucose is firmly established in most people's minds as the prototype of an energy-giving foodstuff, and it is also well known that it can fulfil this function because it liberates energy when oxidized. In point of fact, the free energy change for the complete oxidation of glucose to carbon dioxide and water is about 690 kilocalories per mole. This is a large amount of energy—but, as was said above, it is not immediately utilizable; it must first be converted into the phosphate bond energy of ATP. The usefulness of glucose is limited in just the same way as that of a ten shilling note when confronted with a row of slot machines designed to take pennies; one can do a lot with ten shillings, but only if it is possible to change the note into pennies first.

Getting energy in the form of small change—using the energy of glucose breakdown to build up ATP, that is—must therefore be the guiding principle in the mechanism of sugar breakdown. This mechanism has already been discussed (Chaps. III and IV) in enough detail to indicate its considerable complexity. Cells of the most varied kinds are equipped with the catalytic machinery for performing the whole complicated process—the enzymes to catalyse each of the large number of chemical steps involved.

Now surely, a self-appointed efficiency expert might say, this is a highly inefficient way of running the animal economy, and it is easy to suggest a better one. Let most cells dispense with the cumbersome catalytic machinery required for making ATP at the expense of glucose; and let one organ—the liver, say—specialize in this process. It could then supply "charged" ATP via the blood stream, and other tissues would send back "spent" ADP and phosphate for regeneration. Better still, let the plant kingdom supply ATP direct instead of carbohydrate; animals could excrete ADP and phosphate for recombination by photosynthetic energy, and relieve themselves entirely of the necessity of oxidizing sugars.

A diet of ATP does not sound appetizing, and fortunately there are other arguments against such a drastic reorganization of the system painstakingly devised by natural selection over millions of years. The oxidation of glucose can be conducted in such a way that many molecules of ATP are made per molecule of glucose; probably the maximum number is not far short of 40. In addition, the molecular weight of ATP is more than three times that of glucose. Thus, about a hundred times as much material would have to be shifted around in order to get the same value in terms of energy. It would be just as inconvenient as getting a wage packet paid entirely in pennies instead of banknotes. Glucose may not provide energy in an immediately available form, but it is compact enough for easy and economical transport.

ATP-manufacturing facilities, then, are very necessary equipment for cells. Some ATP is made during the anaerobic phase of breakdown, and more during the aerobic phase. It is easiest to consider the two phases separately.

Getting small change (i) anaerobic phase

Considered from the point of view of ATP synthesis, the reactions of the anaerobic phase (Fig. 11, p. 64) begin disappointingly, for first one and then another molecule of ATP are used up (steps 1 and 3). These early losses may be thought of as a deposit that has to be paid before any change can be obtained. (The deposit is, however, partly returnable, since a proportion of the energy expended is later won back). Later in the reaction sequence, there are two steps (7 and 10) in which ATP is formed. If it is remembered that these steps involve three-carbon substances, and that a single glucose molecule gives rise to two of each, it becomes apparent that a gross total of four molecules of ATP are made; deducting the two molecules used up earlier, the net gain turns out to be two molecules of ATP for each molecule of glucose broken down as far as lactate or alcohol.

The mechanism of ATP formation is here quite clear. The secret lies in the use of phosphorylated intermediates in glucose breakdown. ATP synthesis occurs by transfer of phosphate groups from such intermediates to ADP.

Up to a point, it is enough to know the molecular events by which this occurs; but it is also interesting to enquire further and ask where the necessary energy comes from. Direct combination of phosphate with ADP is, of course, impossible—it is ruled out by the free energy considerations mentioned above; that phosphate as such might combine with ADP is just as unlikely as that water might of itself flow upwards from sea to mountain lake. To effect ATP synthesis, the phosphate must first be raised to a higher energy level, from which it is energetically feasible for condensation with ADP to take place. The general answer to this problem is that the energy required

A COMMON CURRENCY FOR ENERGY TRANSACTIONS

is provided by the transformations undergone by the sugar breakdown intermediates. These transformations, considered in themselves, are energy-yielding; by performing them in combination with phosphate, the opportunity is taken of using the energy they yield to charge the phosphate bond.

As a specific example, the reaction undergone by glyceraldehyde phosphate (step 6) is the most instructive. It might be considered to take place in two stages—the oxidation of the aldehyde to a carboxyl group, and the combination of the latter with inorganic phosphate. The process does not in reality take place by these two steps, which represent a purely theoretical analysis to show the energy relationships involved.

Theoretical stage A $RCHO + H_2O \xrightarrow{-2H} RCOOH$

Theoretical stage B $RCOOH + HO\textcircled{P} \longrightarrow RCOO\textcircled{P} + H_2O$

Sum (reaction 6) $RCHO + HO\textcircled{P} \xrightarrow{-2H} RCOO\textcircled{P}$

The bond binding the phosphate to the carboxyl group is a high energy one; carboxyl-bound phosphate is phosphate raised to a high energy level. This means that, in the subsequent step (reaction 7), it can be transferred to ADP to give ATP. By the same token, however, it also means that reaction B as written can never take place, any more than can the direct combination of inorganic phosphate with ADP. To bring it about requires the input of energy, and this energy is provided by the oxidation of aldehyde to carboxylic acid. In itself, this is an energy-yielding process, and the energy it yields is used to bring phosphate into combination with the carboxyl group. Stage A is the source of power to drive stage B along.

To avoid misunderstanding, it should be emphasized again that the mechanism by which reaction 6 actually occurs is *not* by stages A and B; but this does not alter the basic fact shown by the theoretical analysis—that it is the oxidation of the aldehyde group which provides the energy for the formation of the carboxyl-phosphate bond, and hence ultimately for the synthesis of ATP. In terms of free energy changes of real processes, the situation is this. The oxidation of an aldehyde to a carboxylic acid liberates an appreciable amount of free energy; when the product is a carboxyl-phosphate compound, the amount liberated is less; the difference is the amount put into the carboxyl-phosphate bond.

In step 10, ATP synthesis is effected at the expense of phospho-*enol*-pyruvate, and here again it is made thermodynamically possible by the fact that the phosphate donor has a high energy bond. A useful rule-of-thumb is that a phosphate ester is a high energy one if there is a double bond two atoms away from the phosphorus—that is, if the grouping P—X—Y= is present. The applicability of this rule to ATP, the carboxyl phosphate of 1,3-diphosphoglyceric acid and phospho-*enol*-pyruvate can be seen from the formulae below.

$$\text{adenosine}-\text{O}-\overset{\overset{\text{O}}{\|}}{\underset{\text{OH}}{\text{P}}}-\text{O}-\overset{\overset{\text{O}}{\|}}{\underset{\text{OH}}{\text{P}}}-\text{O}-\overset{\overset{\text{O}}{\|}}{\underset{\text{OH}}{\text{P}}}-\text{OH}$$

$$\begin{array}{c} \text{O}=\text{C}-\text{O}-\overset{\overset{\text{O}}{\|}}{\underset{\text{OH}}{\text{P}}}-\text{OH} \\ | \\ \text{H}-\text{C}-\text{OH} \\ | \\ \text{H}-\text{C}-\text{O}-\overset{\overset{\text{O}}{\|}}{\underset{\text{OH}}{\text{P}}}-\text{OH} \\ | \\ \text{H} \end{array}$$

$$\begin{array}{c} \text{CH}_2 \\ \| \\ \text{C}-\text{O}-\overset{\overset{\text{O}}{\|}}{\underset{\text{OH}}{\text{P}}}-\text{OH} \\ | \\ \text{COOH} \end{array}$$

As the rule correctly predicts, the other phosphate group of 1,3-diphosphoglyceric acid (the one on position 3, at the bottom as written above) is *not* bound by a high energy bond. Esters of phosphate with simple alcoholic hydroxyl groups in general are low energy ones. All the phosphate esters of Fig. 11 except those specifically mentioned above fall into this category; thus ATP is formed at the expense of the only two intermediates on the pathway that are thermodynamically capable of effecting the synthesis.

Getting small change (ii) aerobic phase

The net gain of two molecules of ATP per molecule of glucose broken down anaerobically is, of course, only a small proportion of the total of nearly 40 for complete breakdown as far as carbon dioxide and water. Much the major share of the burden of ATP synthesis is borne by the aerobic phase; and nearly all of it occurs by the process known as "respiratory chain phosphorylation".

What this name indicates is that, as a normal accompaniment of the operation of a respiratory chain, ADP is phosphorylated to give ATP. Clearly, the oxidations supply the energy for the phosphorylation; oxidation and phosphorylation are coupled in some way, so that the energy provided by the one can be used to drive the other (as in the theoretical stages A and B above).

Exactly how this happens is not yet clear, despite a good deal of research that has been devoted to the problem. Even the nature of the various oxido-reduction carriers which make up respiratory chains, it will be remembered, has not been fully elucidated (p. 79). Chemical equations cannot, therefore, yet be written to show by what reactions ATP arises.

It is quite clear, however, that the process does occur and that it is a highly significant one. When freshly isolated mitochondria (p. 75) are supplied with an oxidizable substrate and oxygen, so that a respiratory chain of the type shown on p. 78 operates, ADP and phosphate are at the same time brought into combination. If measurements are made of both the oxygen consumed and of the phosphate esterified, it is found that for every atom of oxygen—for every pair of hydrogen atoms, that is, that is removed from substrate (AH_2, or BH_2, or CH_2) and made to combine with oxygen to form water—up to three phosphate groups can be condensed with ADP. In other words, the maximum P : O ratio is three, to put it in a way that biochemists often do.

Respiratory chain phosphorylation can thus be a very fruitful source of ATP, because many dehydrogenation steps are involved in the complete breakdown of glucose. A rough ATP count can easily be made on the basis that twelve atoms of oxygen are required for complete oxidation:—

$$C_6H_{12}O_6 + 6O_2 = 6CO_2 + 6H_2O$$

With a P : O ratio of three, this gives 36 molecules of phosphate built up into ATP by respiratory chain phosphorylation during complete oxidation of each molecule of glucose. (The calculation is rather over-simplified. One dehydrogenation, that of succinate—step 7 of Fig. 14—uses not NAD or NADP, but a riboflavin derivative; the repiratory chain in this case is rather different, and can esterify only two molecules of phosphate. On the other hand, a small contribution to ATP synthesis is made by a process other than respiratory chain phosphorylation; this occurs in connection with reaction 6 of the citrate cycle, which is in reality more complex than is shown in Fig. 14. The figure 36 thus seems to be quite near the true maximum total for the aerobic phase.)

The mechanism of ATP synthesis, it can be seen, is rather different in the anaerobic and aerobic phases of glucose breakdown. The early stages involve phosphorylated intermediates, from which phosphate groups are transferred to ADP. In the citrate cycle, no such phosphorylated intermediates appear; the reactions of the cycle provide for ATP synthesis mostly in a less direct way, by feeding pairs of hydrogen atoms into respiratory chains, to keep them busily oxidizing and simultaneously phosphorylating ADP.

The loss on the deal—a question of efficiency again

We are now in a position to answer some earlier carpings of the amateur efficiency expert. Why, the question was raised (p. 77), introduce the apparently gratuitous complication of a multi-stage respiratory chain for accomplishing the simple objective of combining hydrogen atoms from substrates with oxygen to form water?

The most telling reply is to flaunt the maximum P : O ratio of three. A single oxido-reduction could hardly be so arranged as to be accompanied by more than a single phosphorylation. Only by interposing a number of intermediate oxido-reduction steps can room be made for as many as three phosphorylations.

Looked at from the energy aspect, the point is this. Reactions of the type

$$AH_2 + \tfrac{1}{2}O_2 \rightarrow A + H_2O$$

in general make available quite a lot of free energy. For typical substrates that feed respiratory chains (e.g., citrate cycle intermediates), the quantity is often of the order of 50 kilocalories per mole. To utilize this free energy efficiently for ATP synthesis, it has to be split into a number of portions each sufficient to form one high energy phosphate bond. This is precisely what the individual stages of respiratory chains do. Instead of being released in one violent bang, so to speak, the energy is divided into manageable little packets which can be more efficiently harnessed to the grand purpose of making ATP—for it is, of course, the energy provided by the oxidations that gives the impetus needed to drive phosphate into combination with ADP.

It bears emphasizing once again that the value of oxidative processes lies not in getting rid of foodstuffs, nor in forming carbon dioxide and water, but in producing energy in a form available for use—which, to a large extent, means high energy phosphate bonds. An animal is not a machine for burning sugar, any more than a car is a machine for burning petrol. Sugar or petrol could be burned far more expeditiously by means other than those employed by animals or cars respectively. If getting rid of petrol were the only object, it could be achieved far more simply and cheaply by putting a lighted match to it than by putting it into a car. The object of a car is to move, and the less petrol it uses in the process, the better is its owner pleased. Similarly, the process of burning foodstuffs is in itself only a liability to an animal (unless it happens to be on a slimming diet, that is). The less foodstuff has to be burned to fulfil its energy requirements, the better for the animal.

The efficiency of biological oxidation, therefore, is most appropriately measured not by the rate at which it makes foodstuffs disappear, but by the proportion of the energy liberated which is actually "caught" in the form of ATP. For the case of glucose, all the relevant figures have already been given. Complete oxidation of a mole of glucose provides 690 kilocalories; when carried out in an animal, it can lead to the formation of up to nearly 40 high energy bonds, each worth 11 to 12 kilocalories. The efficiency works out at 60 to 70%. This, so to speak, is the current rate of exchange for turning banknotes into pennies—six to seven shillings' worth of pennies for every ten shilling note. The 30 to 40% loss on the deal represents the agent's commission for effecting the conversion. By financial standards, this may seem exorbitant, but by the more realistic comparison with man-made machines, the loss seems quite modest. Engineers would be very satisfied with a combustion-powered machine that converts two-thirds of the combustion energy into useful form.

Spending the pennies (i) for chemical synthesis

Having seen how energy is got in the form of immediately available small change, the next question that arises is for what and how it is used. Three main types of objective are recognizable which are vital for the functioning of animal bodies—chemical synthesis, transport of materials by active pumping, and muscular movement. In addition, it is worth saying a little about an entertaining biochemical curiosity, the use of chemical energy to produce light.

Chemical synthesis is, of course, one of the basic functions of life of all kinds. Growth entails synthesis of cellular constituents from smaller molecules. Moreover, synthesis is necessary for maintenance as well as growth. Few constituents of metabolically active tissues just stay put as fixed elements of the metabolic machinery. Most of them undergo quite rapid degradation and must be continuously replaced by synthesis. For many constituents, the half-life—the time, that is, during which 50% is broken down and replaced—is only a few days. (The discovery of this remarkably dynamic state of body constituents was one of the earliest and, to many, the most surprising consequence of the introduction of isotopic methods (p. 102) into biochemical research). It is strange to think that, although adult animal bodies seem to change so slowly with the months, much of them is in fact renewed weekly.

The materials that have to be synthesized for growth and maintenance are very diverse; so, therefore, are the chemical routes that lead to them, and the details of the participation of ATP. This is not the place to plunge too deeply into the chemical particulars of so large a subject. There is one case of synthesis at the expense of ATP, however, the details of which have already been presented, so that it only remains to point out their significance here. The case is that of the synthesis of carbohydrate from lactate.

The route by which this occurs is, in most essentials, the reverse of that shown in Fig. 11. On the breakdown route, it will be remembered, a gross total of four molecules of ATP are formed from each glucose unit. Accordingly, to make the reaction series go in the synthetic direction, four molecules of ATP have to be supplied. (On the other hand, the two molecules of ATP used up in the first and third steps of glucose breakdown are not regenerated during synthesis (cf. p. 106). Although, to achieve synthesis, one has to put in again all that one got out during breakdown, one does not get out again what was put in.) The secret of making ATP during breakdown, it has been pointed out (p. 148), lies in the use of phosphorylated intermediates which donate phosphate to ADP; conversely, ATP is used to effect synthesis by phosphorylating intermediates.

The synthesis of carbohydrate from lactate can be quite an important process for animals. As is well known, muscles under stress form lactate (p. 67). The significance of this is now quite clear—it lies in the ATP formed during anaerobic breakdown. In amount, this ATP is miserably small compared with what could be made by complete oxidative breakdown of the same quantity of carbohydrate, but in the absence of enough oxygen fully to meet a heavy demand, it is the best the muscle can do. When oxygen becomes plentiful again, a small part of the lactate can be oxidized, reaping (by respiratory chain phosphorylation) a rich harvest of ATP which can be used to build the other, larger part of the lactate back into carbohydrate.

In mammals, much of this resynthesis occurs in the liver. Lactate leaks into the blood stream from the hard-pressed muscles, is picked up by the liver and there turned into carbohydrate in the form of the glucose polymer, glycogen. From the liver glycogen, glucose is released into the blood stream as and when necessary to maintain an

A COMMON CURRENCY FOR ENERGY TRANSACTIONS

adequate blood sugar level. Muscles pick up glucose from the blood and turn it into their own reserve of glycogen. This whole series of transformations is sometimes known as the "Cori cycle" (after the two investigators, man and wife, who shared the Nobel prize for medicine in 1947).

The effect of the Cori cycle (Fig. 18) is to make possible a sharing between muscle and liver of the burden and ATP expense of re-synthesizing glycogen from lactate. Paradoxically, some of the muscles' work is thus done in the liver. (A small expenditure of energy, however, still remains for the muscle to bear itself, for the

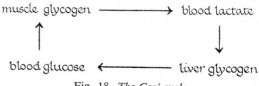

Fig. 18. *The Cori cycle*

conversion of glucose into glycogen is itself a synthesis which demands energy in the form of ATP, involving as it does the first step of Fig. 11.)

Although other specific examples of the use of ATP for synthetic purposes will not be mentioned here (cf., however, p. 177), there is one general point that is worth making before leaving the subject. In the entire discussion of ATP so far, attention has been focussed only on the terminal phosphate bond. The other phosphate-phosphate bond of ATP is, however, equally rich in energy—hence the shorthand way of writing the formula, adenosine—ⓟ∼ⓟ∼ⓟ, where both the "squiggles" stand for high energy bonds. Often only the terminal bond is used, as in the cases discussed above; but it is by no means unknown for both bonds to be used. ATP is then broken down, not merely to ADP, but as far as adenosine monophosphate or AMP; and a double packet of energy is provided for some deserving cause.

Spending the pennies (ii) for pumping

Another process for which animals need energy is the unspectacular but vital one of pumping materials from one place to another. Of course, not all movement of material requires work to be done; in the absence of impermeable barriers, a substance diffuses of its own accord from a region of high to a region of low concentration, until the concentration difference is levelled out. Only for movement in the reverse direction, up a concentration gradient, is active transport with energy input required. Various absorption processes—in the intestine and in the kidney tubules, for instance—are of this kind.

Concentration gradients with respect to cations such as Na^+ and K^+ are important for cells to establish and maintain. As is well known, the major salt in extracellular fluids such as blood plasma is sodium chloride, and cations other than sodium, including potassium, are present in smaller quantities. Inside cells, however, the concentration of potassium far exceeds that of sodium. In human red blood cells, for instance, the concentration of K^+ is 0.15 M, while

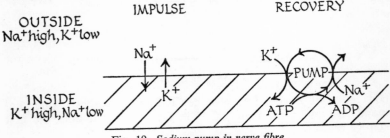

Fig. 19. *Sodium pump in nerve fibre*
Modified from Hodgkin and Keynes.

that of Na^+ is only 0.01 M; in the surrounding plasma, on the other hand, Na^+ at 0·145 M much outweighs K^+ at 0.005 M. The active process by which the difference is maintained consists in pumping sodium out of the cell, and there is good evidence that it requires ATP as its energy source.

An active "sodium pump" is particularly important for nerve cells, for the membrane depolarization which accompanies each passage of a nervous impulse allows sodium to pass in, and consequently potassium to pass out. To keep themselves primed and ready for action, nerve cells must during recovery pump the sodium out again. The dependence of this process on ATP has been shown particularly well with nervous tissue from the squid. In this animal, some of the individual nerve fibres are as much as one millimetre in diameter, and thus easier to experiment with than the much smaller fibres of mammals. Through micro-injection needles, various substances can be injected at will into such giant fibres. When the fibres' own energy-supplying mechanism is poisoned, it can be shown that ATP injection leads to sodium extrusion; and the amount of sodium pumped out varies as the amount of ATP injected.

A COMMON CURRENCY FOR ENERGY TRANSACTIONS

One can form a mental picture, therefore, of an ATP-powered pump which pushes out of the nerve during recovery the sodium that leaked in during the spread of the nervous impulse (Fig. 19). The spread of excitation along a muscle fibre involves similar effects at its membrane, the sarcolemma.

Spending the pennies (iii) for light

Compared to the other, serious purposes to which ATP is put, the production of light may well seem a trivial sideline. The luminescence of fireflies and a few kinds of bacteria could hardly be said to number among the fundamental phenomena of life. It is, nevertheless, quite an amusing form of "light relief", and biochemists have gone some way towards unravelling the processes that enable fireflies to glow in the dark.

An indication of this is the fact that dehydrated firefly tails now figure in the catalogues of biochemical suppliers—probably the most picturesque item in lists that are otherwise dull enough to any but initiated enthusiasts. As publicity, some suppliers even publish photographs taken entirely by firefly light. The exposures are made by the light emitted for a few seconds after ATP is added to suitable firefly extracts.

Further analysis shows that five substances are required for luminescence—a substrate called luciferin, an enzyme called luciferase, magnesium ions, oxygen and ATP. Part of the energy for light production seems to come from the breakdown of ATP, but oxidation of a luciferin derivative is also involved.

Spending the pennies (iv) for muscular work

As a major consumer of ATP, and also as a process of great interest in itself, muscular contraction deserves rather fuller treatment here than other ATP-using processes have been given.

Two proteins that can be extracted from muscle, myosin and actin, have been implicated in the contractile mechanism. They are major constituents of muscle tissue; in the case of rabbit skeletal muscle, which has been studied most, they make up about half the total protein present. Myosin is the one that is present in greater amount—two to three times as much as actin. It has been purified and rather thoroughly studied. Its molecules are very large, with weights of nearly half a million, and have the shapes of long thin filaments—their lengths, at 1,600 Å (10,000,000 Å = 1 mm), being some fifty times as great as their diameters. In 1939, Engelhardt and Ljubimova in Russia made the intriguing observation that myosin acts as an enzyme on ATP, catalysing its hydrolytic breakdown to ADP and phosphate—in other words, that myosin has "ATP-ase" activity. This hydrolysis, of course, can provide energy; release of chemical energy is thus very intimately associated with performance of mechanical work, for one of the two kinds of molecule immediately concerned in causing contraction can itself carry out the hydrolysis.

Of the evidence pointing to myosin and actin as the essentials of the contractile mechanism, the most striking is the fact that threads can be prepared from these two proteins which can be made to contract *in vitro*. When solutions of actin and myosin are mixed, they combine to form a complex called actomyosin, which is soluble in aqueous media only when the salt concentration is fairly high. Squirting an actomyosin solution into water therefore makes the protein complex precipitate as threads of gel. It is only necessary to add a little ATP to the solution bathing the threads to make them contract. ATP concentrations as low as 0·01 to 0·001 M or even lower are enough to bring about the effect; the ATP is broken down meanwhile under the catalytic influence of the myosin component.

How closely can this contraction be compared to that of intact muscle? This is a question which must be pressed as searchingly as possible, for on it hinges the validity of an explanation of muscular action in terms of myosin and actin.

One way in which the two contractions differ is obvious enough to appear even to the naked eye. The contraction of muscle is a matter of becoming shorter and thicker, with barely any volume change; that of actomyosin threads is a matter of shrinkage equally in all directions—"isodimensional" contraction of the gel protein, squeezing out much of the interstitial water. A moment's thought, however, shows that this difference can easily be explained. In muscle, the protein filaments are presumably orientated in the direction of shortening—indeed, there is excellent evidence that they are. (The evidence comes partly from the electron microscope observations described below.) In the actomyosin thread, on the other hand, since it is formed by precipitation from solution, the protein filaments presumably lie randomly in all directions; no wonder, therefore, that contraction here takes place in all directions—this is exactly what must be expected if the filaments all shorten. In favour of this interpretation is the fact that, if a thread of actomyosin gel is first stretched, so that the filaments come to lie predominantly parallel to the long axis, ATP-induced contraction is greater along the length than across the breadth of the thread—"anisodimensional" contraction.

A COMMON CURRENCY FOR ENERGY TRANSACTIONS

Another difference between the two contractile systems lies in the maximum tensions they can develop; actomyosin threads are only about one-twentieth as powerful as intact muscle. The threads are limited in this respect, however, by their low tensile strength—they break before much tension can be developed. If it were possible to give them a breaking strength like that of muscle, they would show up better in the comparison.

These differences, then, need not destroy the hope that the behaviour of the actomyosin gel system with ATP is not merely an amusing artifact but has some genuine physiological relevance. Confidence on this point has been gained by considering contractile systems intermediate in degree of complexity between actomyosin and intact muscle; through a series of such systems, stage by stage comparisons can be made (cf. p. 97) which indicate that the contractile process is fundamentally similar in all of them. Among the intermediate systems that have been studied, the most notable are isolated myofibrils and suitably treated individual muscle fibres.

Myofibrils are microscopically visible contractile elements with diameters of the order of 1 μ (one-thousandth of a millimetre), many of which run along the length of each muscle fibre, held together by the fibre membrane, the sarcolemma. Muscle fibres are highly specialized, multinucleate cells. The fluid they contain, filling the space between the myofibrils, is called the sarcoplasm and has the composition of an ordinary intracellular fluid, containing various salts, those of potassium predominating (p. 156); dissolved in it are also present the various enzymes and coenzymes required for the anaerobic breakdown of glycogen to lactate, so that it is capable of supplying some ATP. Further ATP requirements are catered for by mitochondria lying between the myofibrils.

By mechanically breaking muscle tissue, myofibrils can be isolated from it and shown to be composed largely of myosin and actin. If a suspension of myofibrils under a microscope is irrigated with 0·001 M ATP solution they can be seen to contract, reaching little more than half their original lengths; this contraction is like that of intact muscle in that shortening is accompanied by thickening. At the same time, of course, ATP is hydrolysed by the myosin present.

When thin pieces of muscle are treated at low temperatures with glycerol-water mixtures, nearly all the low molecular weight material and a good deal of the sarcoplasmic protein is extracted; what remains is largely the myofibrillar framework composed of myosin and actin—virtually the contractile apparatus of the fibres in isolation, not disorganized but stripped of much of the accessory machinery that normally accompanies it. Here again, dilute solutions of ATP cause contraction. The ATP-induced contraction of single fibres from such preparations can be studied in a quantitative way, so that exact comparisons with the behaviour of living muscle are possible. Notably, it has been found that tensions of up to 4,000 g per square centimetre of cross-sectional area can be developed; this is of the same order as the maximum tension reached by muscles of warm-blooded animals.

There are good grounds, then, for taking the effect of ATP on actomyosin gel to be the prototype, on the molecular level of organization, of the contraction of intact muscle. It seems to be essentially a real biological event isolated—standing in a relation to physiological contraction rather like that in which the clotting of fibrinogen by thrombin stands to the clotting of blood (p. 53). The relation is more complicated in the case of muscle, however, for organization above the molecular level is important here. Fibrinogen, after all, is present in free solution in blood, whereas myosin and actin are arranged in very definite ways in muscle.

Contraction is thus an altogether more difficult problem than clotting, for biochemical explanation here demands correlations between the molecular and tissue levels. This is just the type of correlation which has been presented, in a general context, as perhaps the toughest challenge facing biochemistry (p. 97). In many cases, it is hardly possible to take up the challenge at all. In the particular case of muscle, fortunately, it can be answered in a relatively satisfactory way. Although understanding certainly remains far from complete, it is possible in this instance to offer a reasonably plausible and coherent picture of the way in which molecular events are organized to give rise to the phenomena observed at tissue level. Various forms of microscopy have contributed crucial observations to make this possible.

The characteristic cross-striated appearance of skeletal muscle under the ordinary light microscope is due to the myofibrils, which themselves have a banded structure, the bands of adjacent myofibrils being accurately aligned. The bands are called the A and I bands (for anisotropic and isotropic, according to whether they are or are not doubly refracting). In the middle of each I band is a Z disc, and the functional segments of the myofibrils are best thought of as extending from one Z disc to the next; these segments are called sarcomeres. The middle portion of each A band presents a slightly less dense appearance and is known as the H zone (after the microscopist Hensen) (Fig. 20).

These are the features which were recognized by classical microscopy. Since the mid-twentieth century it has been possible, by means of the newer techniques of interference and phase contrast microscopy, to settle the hitherto vexed question of the changes in the bands during shortening. Furthermore, electron microscopy has suggested an elegant interpretation of the observations in molecular terms.

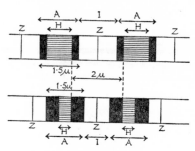

Fig. 20.

Myofibril, extended and contracted, showing the two distances which remain constant. (This refers to moderate contraction; in extreme contraction, the A bands meet and new bands are formed at the positions of the Z discs.)

It is the I band which shortens in contraction. The A band remains the same length, but the H zone contracts so that it takes up less of the A band. Two distances stay constant during both contraction and stretch—the A band remains $1 \cdot 5\,\mu$ long, and the inter-H distance, from the end of one H zone to the beginning of the next, remains $2\,\mu$ (Fig. 20).

The simplest interpretation of the two distances that remain constant is that they represent rods which do not themselves shorten but cause shortening by sliding past each other during contraction. Two kinds of rod are actually visible in electron microscope photographs, one of them considerably thicker than the other. The two sets partially overlap and interdigitate as shown diagrammatically in Fig. 21. A transverse section through the dense part of an A band shows both kinds of rod, in strikingly regular hexagonal array. In the H zone only the thick, and in the I band only the thin kind is seen. During contraction, the two sets are imagined to slide into each other so that the degree of overlap increases.

The thick rods can be shown to be made of myosin and the thin of actin. The relative proportions are about right for this identification—two to three parts by weight of myosin to one of actin; and extraction procedures known to extract myosin or actin selectively remove the material of the thick and the thin rods respectively.

We are back, therefore, to the molecules that can be studied in purified form in the test tube. The story of muscle can be taken as a case-history—perhaps as illuminating as any that is available in the present state of knowledge—of how biochemistry can set out to explain directly observable events in living organisms. Between

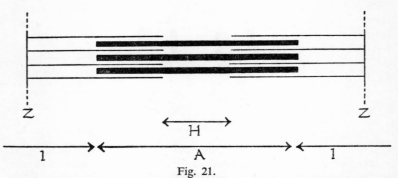

Fig. 21.
Diagrammatic representation of the structure of a myofibril as seen in longitudinal section by means of the electron microscope.

isolated molecules and intact muscles, a chain of critical comparisons is set up through systems of intermediate degrees of complexity (cf. p. 97). Certain vital links in this chain, falling as they do between the molecular and cellular levels of organization, have only relatively recently become accessible to investigation. On the one hand, studies of isolated ATP, myosin and actin have been extended through actomyosin gel to myofibrils, which are specialized sub-cellular structures, and to the still more highly organized myofibrillar framework remaining in fibres after extraction with glycerol-water mixtures. On the other hand, newer types of microscopy have provided vital data on fine structure.

Some sort of a bridge of continuity has thus been set up between the molecular and the tissue levels of organization, which makes it possible to examine whether and how studies on the one level are relevant to events on the other. The bridge has been formed by a link-up between extensions of two classical approaches—the chemist's aspiration towards more complex molecular systems, the biologist's towards finer microscopic resolution (cf. p. 109). The link-up is not, of course, complete and final—but at least it can be visualized in fair detail. The rods visible in electron microscope pictures are not individual molecules—but they are only one order of magnitude greater; the length of an A band is only about ten times that of a myosin molecule. How exactly the mental picture of sliding rods in myofibrils fits in with the ATP-induced contraction of actomyosin gel cannot be said in detail—but it is not impossible to imagine answers. The mere fact that the questions can be put in such intimate detail is a gratifying achievement. Those who peer down microscopes and those who wield test tubes, instead of each talking about their own problems in their own terms, have at least reached a common agenda on which to confer.

IX

Transmitting Information—Biochemical Genetics

Miniature or message? The preformationist and modern views compared

What is it that passes from generation to generation to make offspring resemble parent? This is surely one of the most basic questions of biology, and one to which many different kinds of answer have been given. In past centuries, biologists expended considerable effort on naming and defining metaphysical life principles capable of guiding growth and differentiation. To us, this may seem useless enough—but it was more sensible than some of the proposals that have been made regarding a physical basis for heredity. Notable among these was the theory that acquired the label "preformation". From the vantage point of the twentieth century, it is so easy to pour scorn on the mental aberrations of earlier ages that they are hardly fair game; but even the utmost historical sympathy leaves room for wonder at the extremes of absurdity to which the theory of preformation was pushed. Yet it was widely accepted for much of the relatively recent and intellectually mature eighteenth century.

According to the preformationists, the development of an individual does not involve formation of new bodily structures, but only increase in size of previously existing ones—"an unfolding of what was already there, like a Japanese flower in water", as Needham, himself a biochemist as well as a historian of science, puts it. Partly, the rise of this theory can be traced to an unlucky accident. It happened to be in the heat of an Italian August that Malpighi, the great seventeenth century biologist, made his study of the development of the chick. His observations, made with a simple microscope, were in many ways admirable, but the high temperature may well have brought the eggs quickly to an unusually advanced stage, and unfortunately he did not bother to examine eggs before laying. Thus he could never find an absolutely undeveloped germinal spot, and concluded that "we see an emerging manifestation of parts successively, but never the first origin of any of them".

Philosophers found the idea intriguing. The French priest Malebranche wrote: "We must suppose that all the bodies of men and animals which will ever be born until the consummation of time will have been direct products of the original creation—in other words, that the first females were created with all the subsequent individuals of their own species within them". The Dutchman Swammerdam, himself an accomplished microscopist, but with mystical leanings, saw in preformation an explanation of original sin. "In nature," he wrote, "there is no generation but only the growth of parts. Thus original sin is explained, for all men were contained in the organs of Adam and Eve. When their stock of eggs is exhausted, the human race will cease to be".

By the beginning of the eighteenth century, the doctrine of preformation was rather thoroughly established, and the main cause of controversy was whether the preformed miniature embryos are in the eggs of the females or in the spermatozoa of the males. Support for the latter view came from over-enthusiastic microscopists who claimed to have seen minute forms of men, complete with arms, legs and heads, inside human spermatozoa; one (Gautier) even published a drawing of a microscopic horse in equine semen, and noted the very large ears of the corresponding animalcule in the semen of a donkey.

Microscopists may easily be led astray by vivid imagination—that has happened in more recent times too. The really absurd thing about the preformation doctrine was the immense number of eggs or seeds with which ancestral animals had to be endowed. In 1722, the calculation was made that, even on the naïve interpretation of the Bible that put the age of the world at under 6,000 years, the number of rabbits in the first rabbit must have been something like a 100,000-figure number. Even this consideration was not enough to deter many leading biologists of the time.

As microscopists acquired better instruments, and more rigorous intellectual discipline, talk of animalcules was replaced by talk of nuclei with their chromosomes. Geneticists developed the idea of hereditary units or genes linearly arranged on the chromosomes. Only quite recently—since around 1940—has it been possible to push the analysis a stage further, down to the level of molecules. This, of course, is precisely what biochemists want to do for as many biological phenomena as possible. It is an advance on the chromosome concept in another way, too, because it embraces the large number of micro-organisms which lack a morphologically well-defined apparatus of nuclei and chromosomes but nevertheless manage to pass on hereditary characteristics to their offspring. (Much work has, in fact, been done on micro-organisms, which offer the great convenience, as experimental material, of being able to produce many generations between start and end of a research worker's day.)

It is now believed, with excellent experimental justification, that the genetic material has the chemical nature of deoxyribonucleic acid, called DNA for short. (In living cells, this exists in the salt form, of course, so that it should really be called deoxyribonucleate, but the generally used abbreviation is formed from the initials of deoxyribo-nucleic acid.) In the structure of DNA lie the factors which determine the heredity of the offspring. The form of the future adult is contained in it—not, however, merely compressed into a small space, but also translated into a different kind of shape. It is like the written description of an object, made up of shapes not resembling it and yet capable of representing it to a reader. A parent does not produce a tiny embryo which has merely to grow bigger, but a set of precise descriptive instructions on how to make an embryo. The instructions are embodied in a linear code, like a tape with writing in morse-alphabet on it; they have to be decoded or read and put into effect before a shape emerges that is recognizable as an embryonic living organism. Furthermore, since it is absurd to suppose that all the genetic instructions at present in existence have existed since the beginning of the world, there must be a good mechanism for copying or duplicating the instructions with substantial accuracy.

In essence, then, the difference between the modern and the preformationist views is that it is a message, not a miniature, that passes from parent to offspring. A broad significance can be detected in this, for it is easier to duplicate messages. The information in a message can be conveyed in essentially one or two dimensions, which makes it easier to reproduce than the three-dimensional object it describes—just as the instructions and plans for building a house are easier to reproduce than miniature models of the house.

The basic objectives of biochemical genetics now appear more clearly. They must include knowledge not only of the structure of the heredity-carrying material, but also of the way in which it is duplicated and of the way in which the information contained in it reaches its expression in developed living organisms. The discussion which follows takes up these topics in turn.

The bearer of the message—DNA

DNA, the bearer of genetic information, can be isolated as very large molecules, with molecular weights as high as ten million. It is

TRANSMITTING INFORMATION—BIOCHEMICAL GENETICS

polymeric in construction, and on hydrolysis yields constituents of three types—phosphate, sugar and nitrogenous bases. The phosphate residues give DNA its acidic character. The sugar is deoxyribose, which is the five-carbon sugar ribose minus the oxygen at carbon atom number two. It is this constituent which makes "deoxyribonucleic acid" into the heavy-weight name that it is; the qualification is necessary because another kind of nucleic acid contains ribose and is distinguished by the name ribonucleic acid or RNA.

```
1    CHO              CHO
     |                |
2    HCOH             CH₂
     |                |
3    HCOH             HCOH
     |                |
4    HCOH             HCOH
     |                |
5    CH₂OH            CH₂OH
     D-ribose         2-Deoxy-D-ribose
```

Four different nitrogenous bases are present in DNA—cytosine, thymine, adenine and guanine. Two of these contain the pyrimidine nucleus, a six-membered ring with two nitrogen atoms; the other two contain the purine nucleus, which has in addition another ring, this time five-membered but also with two nitrogen atoms. The positions can be numbered as follows:

```
        6                    6       7
        C                    C       N
      /   \                /   \ 5 /
    1N     C5            1N     C      \
    |      |             |      |       C8
    2C     C4            2C     C      /
      \   /                \   / 4 \ /
        N                    N       N
        3                    3       9
  Pyrimidine nucleus       Purine nucleus
```

The differences between the individual bases lie in the substituents attached to these ring systems. In looking at them, it is worth focussing attention at once on those in the 6 positions (at the top, as written below) and noting that in both the pyrimidine and purine pairs, one has an amino group here, the other a carbonyl group.

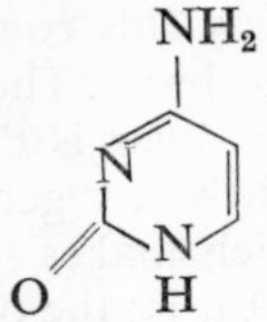

THE BIOCHEMICAL APPROACH TO LIFE

Cytosine

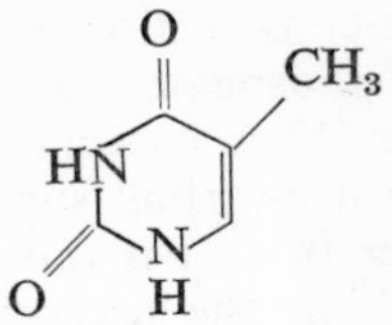

Thymine

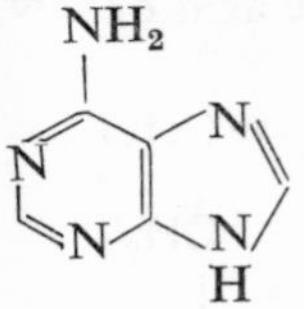

Adenine

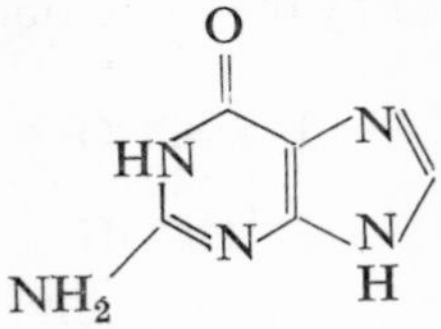

Guanine

The repeating units of which DNA is made are called nucleotides; in them, the constituents are combined in the order base-sugar-phosphate. Since the sugar and the phosphate are the same in all the nucleotides of DNA, there are four kinds of them, according to the nitrogenous base present. They are represented below as C, T, A and G — C standing for cytosine-deoxyribose-phosphate, T for thymine-deoxyribose-phosphate, and so on. Linkage between the nucleotides is between the phosphate and sugar residues, so that a nucleic acid can be represented as:

Base—Sugar
　　　＼
　　　　Phosphate
　　　／
Base—Sugar
　　　＼
　　　　Phosphate
　　　／
Base—Sugar
　　　＼
　　　　Phosphate
　　　／

This structure is best thought of as a sugar-phosphate chain, with various bases sticking out from it.

An important aspect of nucleic acid chemistry is the determination of the ratios of the various bases present. Much of biochemistry,

TRANSMITTING INFORMATION—BIOCHEMICAL GENETICS

naturally, is dependent on the sensitivity and precision of analytical techniques; but seldom have results of such momentous theoretical interest come from refinement of analytical values as in the case of the base contents of nucleic acids. Old, relatively inaccurate values showed roughly equimolar amounts of the four bases to be present, and led to the "polytetranucleotide hypothesis" of nucleic acid structure. According to this, the nucleotides are arranged in repeating quartets in which each kind of base is represented in fixed order, mere repetition of the tetranucleotide groups forming long chains. Such a structure admits of far too little variation for an adequate genetic code to be written in it. The many specific characters which can be hereditarily transmitted could not be represented with such limited means; and DNA could not, while the polytetranucleotide hypothesis was current, be considered a likely candidate for the genetic material.

Modern chromatographic methods have shown, however, that the molar base ratios often differ quite widely from the 1 : 1 : 1 : 1 ratio required by the hypothesis, which has therefore been abandoned. Without the restrictive grouping into repeating units of four, the nucleotides can be arranged in a vast number of different sequences. The number of possibilities, in a chain of n nucleotides of four different kinds, is 4^n (cf. p. 30). By the time n reaches 100, 4^n has already exceeded the number of atoms in the solar system. DNA molecules contain thousands of nucleotides, so that the number of possibilities is much greater still.

The four different nucleotides, A, G, C and T, can now be thought of as making up a four-letter alphabet. By flexible rearrangements of the four symbols, a great variety of complex messages can be represented. Four different symbols may not at first sight seem much for an alphabet, but to see how ample they are, one has only to think again of the morse-alphabet, which manages with only two. With nothing other than arrangements of dots and dashes, there is no difficulty in conveying all the information contained in the Bible, the works of Shakespeare or a voluminous text-book of biochemistry.

Identifying the bearer—evidence for the genetic role of DNA

Obviously, the range of variation made possible by a flexible DNA structure does no more than to make it plausible that DNA is the carrier of genetic information; it does not prove that this is really so. There are, however, good grounds for the belief that it is. Some of the most convincing ones will emerge later, but some of the simpler ones may be mentioned now.

As has been well known for some time, there is ample evidence from cytological work that hereditary characters are carried in the nuclei rather than the cytoplasm, and more specifically in the chromosomes of the nuclei. DNA in cells is confined to the nuclei. (RNA, by contrast, is not; most of the RNA of cells is in the cytoplasm, so that the name ribonucleic acid is something of a misnomer.) Moreover, the amount of DNA per nucleus is nearly constant for the various somatic tissues of a given species, though it may be quite different in another species. In the rat, for instance, the amount is 6·5 to 6·7 picograms (1 picogram = 10^{-12} g) per nucleus whether kidney, spleen, lung, leucocytes or heart are studied; in the fowl, on the other hand, the figure is 2·3 to 2·6 picograms. This correlates well with the fact that these somatic cells all have the same chromosome complement in a given species, while other species have quite different chromosome constitutions.

Two exceptions to the rule that DNA per nucleus is constant further strengthen the correlation with chromosomes. Germ cells are haploid—that is, they contain only half the somatic chromosome complement; in their formation, a meiotic division is involved which halves the chromosome number, so that when egg and sperm cell fuse the somatic or diploid number is restored. Sure enough, the amount of DNA per sperm cell is found to be half the amount characteristic of the somatic cells of the species. Conversely, some tissues have polyploid nuclei with two or four times the normal chromosome number; rat liver is such a case, and here is found a good deal more DNA per nucleus than the 6·5 to 6·7 picograms that is otherwise standard for the rat.

Perhaps the best evidence that DNA is genetic material comes from studies with bacteria of the phenomenon called "transformation". The classical instance of this is observed with pneumococci. These bacteria ordinarily surround themselves with a capsule made of polysaccharide, but a kind is also known which makes no such capsule. Ability to produce a capsule is a hereditary characteristic, just as ability to make a pigment might be in a plant or animal. If a suitable extract of the capsulated type is added to a medium in which non-capsulated organisms are growing, some of them acquire the ability to make a capsule, and can, furthermore, hand this ability on to their progeny for many generations—they have been genetically "transformed", in fact. The transforming agent in the extracts has been isolated and shown to be DNA. Some of the added DNA clearly gets into the recipient cells and carries with it the genetic information necessary to make a capsule.

Other bacteria can be transformed with respect to other hereditary traits.

Copying the message—the duplication of DNA

Although the newer and better analytical data for the base composition of DNA wrecked the simple 1 : 1 : 1 : 1 ratio of the old polytetranucleotide hypothesis, they did at the same time bring to light other equivalences. These are even more intriguing, and contribute an essential item to the present picture of the mechanism by which DNA duplicates itself.

What the analytical data show is that for every residue of adenine there is one of thymine, and that the guanine and cytosine contents are likewise equivalent. Like the amount of DNA per nucleus, its base composition is the same in various tissues of any one species, but different in other species. From whatever species DNA is isolated, however, adenine and thymine are found in equimolar amounts, and the same applies to guanine and cytosine. Usually, adenine preponderates over guanine, and there is correspondingly more thymine than cytosine. In some microbial DNA's, this situation is reversed. In all cases, however, $A = T$ and $G = C$. (One qualification must be added. Certain cytosine derivatives may partially or entirely replace cytosine itself.) As corollaries of these equivalences, it may be noted, the sum of the purines equals the sum of the pyrimidines, and the sum of the amino substituents at position 6 equals the sum of the carbonyl groups at that position.

Our present picture of the three-dimensional configuration of DNA takes into account the observed equivalences between pairs of bases. It was proposed in 1953 by Watson and Crick at Cambridge, and was based also on data obtained by X-ray diffraction (p. 33) and on evidence that hydrogen bonds are important in maintaining the configuration of DNA, as they are in the case of proteins (p. 22). The suggested structure takes the form of a double helix of sugar-phosphate chains; two such chains are intertwined and coiled about a common axis. The bases are on the inside of the helix, stacked perpendicular to the axis, and hydrogen bonding between pairs of them holds the two chains together (Fig. 22).

What makes the structure so intriguing is that the bases will fit into it only in specific pairs. In Fig. 22, thin rods represent the base pairs through which the two sugar-phosphate strands are hydrogen-bonded to each other, but in fact the atoms of the purines and pyrimidines fill the inner space of the helix rather completely. By making scale models, Watson and Crick showed that a purine contributed by one chain must always go opposite a pyrimidine contributed by the other; there is simply not enough room for a purine to go opposite a purine. When the requirement for hydrogen bonding was further taken into account, it emerged that adenine must always go

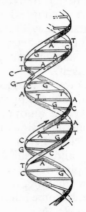

```
   \Sugar——Base-------Base——Sugar/
Phosphate                    Phosphate
   /Sugar——Base-------Base——Sugar\
Phosphate                    Phosphate
   \Sugar——Base-------Base——Sugar/
```

Fig. 22. *Watson-Crick model of DNA*

The spirals represent two sugar-phosphate chains; the rods indicate how they are held together by hydrogen bonding between pairs of bases, as shown by the dotted lines in the scheme.

opposite thymine, and guanine opposite cytosine (Fig. 23). The base sequence of one chain must therefore be complementary to that of the other. If a segment of one chain is . . . ATTCG . . . , for instance, the chain opposite must be . . . TAAGC . . . .

This model of two complementary coils does more than to offer an explanation of the observed regularities in the base composition of DNA—it also suggests a means by which accuracy of duplication may be ensured. If the two chains of a helix are separated and a new partner for either or both is synthesized opposite the old chain from

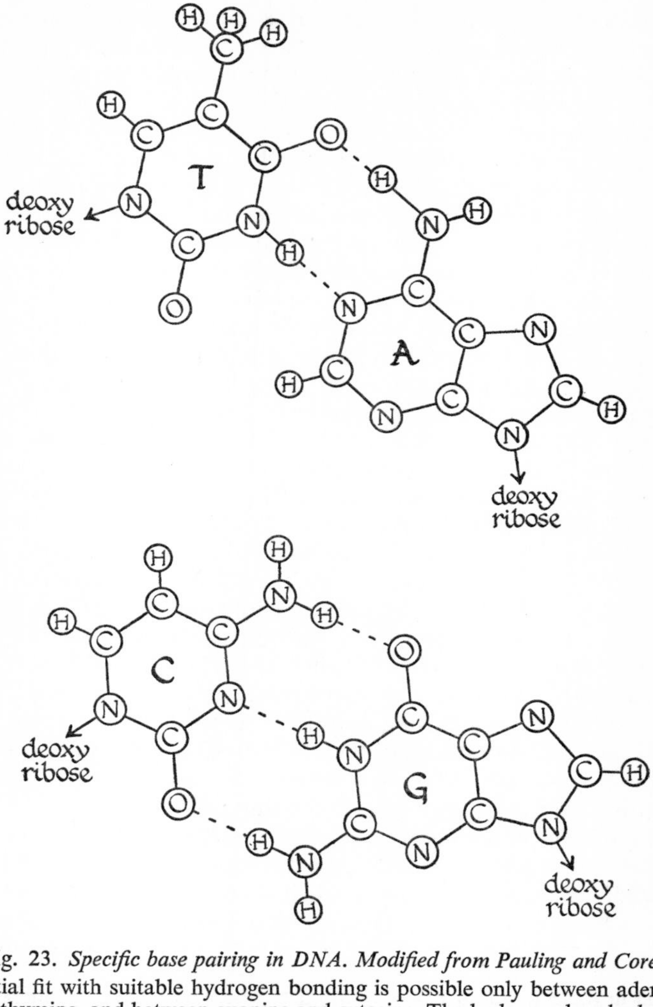

Fig. 23. *Specific base pairing in DNA. Modified from Pauling and Corey*
Spatial fit with suitable hydrogen bonding is possible only between adenine and thymine, and between guanine and cytosine. The hydrogen bonds shown uppermost in each base pair are between 6-amino and 6-carbonyl groups, so that the suggested pairing offers a structural basis for the observed equivalence between these groups as well as that between purines and pyrimidines.

A simple view of a hydrogen bond is that it is formed by an H atom between two electronegative atoms like N or O; bonding occurs because the H nucleus associates itself to some extent with the atom other than the one to which it is nucleotides or nucleotide derivatives, the characteristic base sequence is preserved. The old strand acts as a mould or template which determines the form of the new one. Such a mechanism could account for the way in which genetic information is passed on substantially unchanged from generation to generation.

In its details, the Watson-Crick model remains only a hypothesis, but it has been widely and enthusiastically accepted and its proposers shared the Nobel prize for medicine with Wilkins in 1962. The general idea of two complementary strands has received a good deal of experimental support. Perhaps the most elegant comes from an experiment done by Meselson and Stahl in California in 1958. It relied on the use of the heavy isotope of nitrogen, ^{15}N, which differs from ordinary nitrogen in having a mass fifteen instead of fourteen times as great as that of a hydrogen atom (p. 102). Meselson and Stahl grew the common intestinal bacterium *Escherichia coli* in a medium in which the entire nitrogen supply was in the form of ^{15}N. The DNA formed, because of its ^{15}N content, was appreciably denser than normal, ^{14}N-containing DNA, and the two varieties of DNA could be separately detected by a refinement of the technique of ultracentrifugation (p. 50). Once fully labelled with ^{15}N, the cells were transferred to a medium containing compounds of ordinary ^{14}N, and allowed to reproduce for several generations, DNA being isolated and tested by ultracentrifugation at various stages.

The results were in accordance with the predictions of the theory (Fig. 24). After one generation, when the DNA should have duplicated once, it had a density intermediate between that containing only ^{15}N and that containing only ^{14}N—as would be expected if its helices each had one strand of the old ^{15}N—DNA and one strand of the new ^{14}N —DNA. After two generations, the same 50 : 50 DNA was still found, but now together with normal ^{14}N-DNA, both strands of which had been synthesized since the transfer to ^{14}N medium. In succeeding generations, the same two varieties of DNA persisted, the amount of the half-labelled one progressively diminishing in amount compared to the normal unlabelled one. This pattern

---

Notes to Fig. 23.
  joined in the conventional valency formula. Thus the H atom of an >NH group attaches itself partly to the oxygen of a suitably placed carbonyl group.

$$>N-H---O=C<$$

Three of the five hydrogen bonds shown in the figure are of this type, and it is worth noticing that ample opportunities for similar bonds in proteins (p. 22) are provided by the profusion of —CO— and —NH— groups all along the peptide chains.

of DNA distribution to progeny is good evidence for a double-strand mechanism, parent strands persisting intact and acting as templates for making new strands.

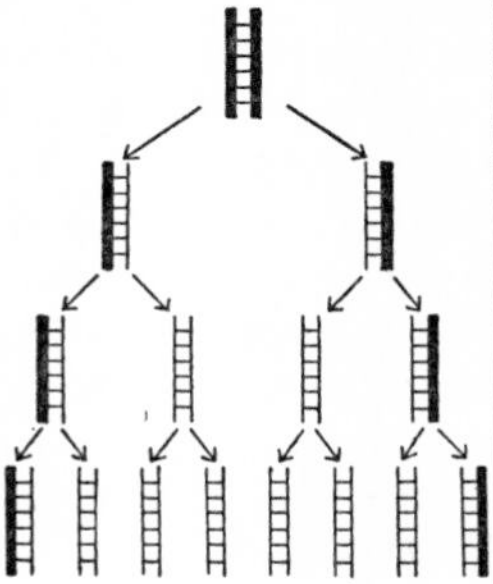

Generation	Proportion of DNA		
	Fully labelled	Half labelled	Not labelled
Parent	All	0	0
1st	0	All	0
2nd	0	½	½
3rd	0	¼	¾

Fig. 24.

Predicted distribution of DNA to progeny in the Meselson-Stahl experiment, assuming a double-strand mechanism. The thick lines represent strands of "heavy" DNA labelled with ^{15}N, the thin lines strands of normal ^{14}N-DNA.

Decoding the message—the control of protein synthesis by nucleic acids

If it is true that DNA carries the genetic message from parent to offspring, it still remains to be determined how the instructions in the message are put into effect. Having seen how prominent a place in the biochemical scheme of things is occupied by the proteins, particularly in their roles as enzymes (pp. 52, 70), it is not surprising that the answer seems to be (in part at least) that DNA controls the structure, and hence the activity, of the proteins made by cells. Somehow, the sequence of nucleotides in DNA determines the sequence of amino-acids in the protein synthesized.

Proteins are built of about 20 different sub-units (p. 19), nucleic acids of four. If a nucleotide sequence is to determine an amino-acid sequence, therefore, a single amino-acid must be represented by a group of nucleotides. Probably a characteristic set of three nucleotides stands for each amino-acid. The sequence AGA, for instance, might stand for one amino-acid, the sequence GCC for another. Four different nucleotides give an ample number of different triplets to provide a characteristic one for each amino-acid, for the total number possible is 4^3 or 64. The morse code is again a useful analogy to the nucleotide sequence. Translating morse code into ordinary language involves turning dots and dashes in characteristic groups of three into the 26 letters of the ordinary alphabet; similarly, turning a nucleotide sequence into an amino-acid sequence means "reading" a particular amino-acid for each characteristic nucleotide triplet, the four-letter nucleotide alphabet becoming the twenty-letter amino-acid alphabet. The coded nucleotide message is thereby decoded into the more workaday language of proteins.

It would be easy to visualize how such a procedure operates if all the protein synthesis in nucleated cells took place in the nuclei, where the DNA is. The amino-acids, one could imagine, find their appropriate triplet on the nucleic acid chain and thereby assemble themselves in a determined order; some "zipper" mechanism then comes along to join them by peptide bonds into a long peptide chain. The nucleic acid would be acting as a template shaping the synthesis of a protein chain, more or less as it does for the synthesis of a new nucleic acid chain. Unfortunately for this simple view, much of the protein synthesis in cells occurs outside the nucleus. The principal protein assembly sites are the cytoplasmic particles called ribosomes (p. 72). Something presumably carries information from the nucleus out to the ribosomes. It appears that this function is carried out by a small but active fraction of the cell's RNA, the "messenger RNA".

In chemical ground-plan, RNA is like DNA except in that ribose takes the place of deoxyribose (p. 167), and that one of the bases, thymine, is replaced by a different pyrimidine, uracil. (One of the nucleotides of RNA, it may be noted, is adenine-ribose-phosphate; this grouping also occurs in ATP (p. 64), the adenine-ribose portion making up adenosine. There may well be more significance in this chemical coincidence than can as yet be fully appreciated.) Since RNA resembles DNA in having four different bases, it could form a similar four-letter alphabet. It seems plausible that the DNA four-letter code is first transcribed into an RNA four-letter code, and that this in turn is translated into the twenty-letter amino-acid alphabet. DNA in the nucleus seems to direct the synthesis of specific RNA, which passes out to the ribosomes to direct the synthesis of protein. While the master-plan is in the form of DNA, RNA molecules could be the blue-prints used as working copies at the actual manufacturing sites.

The experimental evidence in favour of a template role for RNA in protein synthesis is now impressive. One notable study, made by Nirenberg and Matthaei in the United States in 1961, has shown that it is possible to make protein-synthesizing machinery work according to the instructions of an artificial RNA message. The system used in this investigation consisted of two fractions obtained by differentially centrifuging broken cells of the bacterium *Escherichia coli*—the ribosomes and the non-particulate fraction or cell sap left as supernatant after centrifuging down ribosomes (p. 75). These two fractions together, supplied with a mixture of the twenty naturally occurring amino-acids, were capable of synthesizing protein. (As a source of energy for the synthesis, ATP also had to be supplied (p. 155). Roughly speaking, the cell sap contains the catalytic apparatus for making amino-acids react with ATP and thereby "priming" or "activating" them with enough energy to make the synthesis proceed; in the ribosomes, the "activated" amino-acids are assembled into proteins.)

When the RNA present was destroyed by adding a trace of an enzyme that hydrolyses it (an "RNA-ase"), protein synthesis was entirely abolished. Adding suitable RNA preparations, on the other hand, stimulated the synthetic activity of the intact system under certain conditions. More notable still was the result obtained by adding an artificially synthesized polymer of the nucleotide uracil-ribose-phosphate—an unnatural prototype of RNA in which, instead of the four bases of normal RNA, only uracil is present. Added to the *E. coli* system, this polynucleotide stimulated synthesis, but the product was no ordinary protein; a polymer containing only the amino-acid phenylalanine was formed. The unnatural template, in other words, had induced the synthesis of the unnatural "protein" polyphenylalanine.

This result not only gives direct evidence that a nucleotide sequence can determine an amino-acid sequence—it also gives some information about what the RNA code is. Phenylalanine, it seems, can be represented in RNA by the letter U only (where U stands for uracil-ribose-phosphate). Presumably, if it is always a triplet of nucleotides that stands for each amino-acid, the code word for phenylalanine is UUU. By experiments along these and other lines, it seems likely that the RNA code will soon be cracked by the identification of all the nucleotide triplets corresponding to amino-acids. The whole region of phenomena concerned with nucleic acids and protein synthesis has, since mid-century, formed one of the most significant growing-points of biochemistry, and further exciting developments can be expected here in the near future.

The wrong messages—viruses

Of the various lines of experimental evidence that both types of nucleic acid do in some way play a part in directing protein synthesis, some of the most striking comes from work with viruses. All viruses contain protein and nucleic acid, either RNA or DNA, and often little or no other material is present. When the nucleic acid enters a host cell, it diverts the synthetic machinery from its normal purpose of making cell protein, and puts it to work instead at making virus protein. The virus nucleic acid provides a new set of instructions for manufacture—a wrong set, from the host cell's point of view.

A case that has been particularly well studied is that of tobacco mosaic virus. Like other plant viruses, this one contains protein and RNA. It has been possible to separate the RNA from the protein by mild treatment with detergents, and to show that it can by itself infect tobacco plants. Once inside a plant cell, the virus RNA makes a successful take-over bid for its protein-manufacturing capacity. Protein synthesis continues, but it is largely virus instead of host protein that is now made. The protein-synthesizing mechanism now operates according to the instructions of the virus RNA instead of the host's own characteristic RNA; a new blue-print has been supplied, and a new type of protein comes off the assembly lines. (Virus RNA is made too, of course, so that more whole virus is formed.)

Another case about which relatively detailed information is available is that of a bacterial virus or bacteriophage which attacks the bacterium *Escherichia coli*. Here, the nucleic acid is DNA; when a bacterial cell is infected, the DNA enters, leaving nearly all the protein outside. The metabolic machinery of the host cell now becomes harnessed to the synthesis of bacteriophage instead of bacterial constituents, so that phage protein and phage DNA appear in quantity. There is reason to believe that the infecting phage DNA first directs the formation of a characteristic RNA, which in turn supervises protein synthesis in such a way that phage-specific protein results.

Putting the instructions into effect—genes and enzymes

Biochemists tend to think that a living organism is what its proteins make it. To the extent that this is so, the genetic message can determine its hereditary characters as a whole if it can determine its protein make-up.

It bears repeating that all enzymes are proteins, and specific catalysis by enzymes determines what chemical changes shall occur (pp. 80, 146). Synthesis of biological materials is therefore under the immediate directive control of enzymes. This gives proteins a key position, for they can determine the synthesis of other body constituents. It is thus quite easy to see how protein make-up can influence some aspects of visible form. Presence or absence of a pigment, for instance, may depend on whether or not one of the enzymes concerned in its synthesis is properly made. It is more difficult, however, to form a mental picture of how protein make-up could determine anatomical characteristics, such as five-fingeredness, or the more detailed features that make fond aunts exclaim how like his father a squalling baby is. Perhaps it is wise to bear in mind that it remains no more than a hypothesis that the genetic information expresses itself *only* through protein structure, and in no other way.

It can, nevertheless, be taken as proved beyond reasonable doubt that *some* hereditary characteristics depend on the synthesis of individual proteins. Studies bearing on this matter largely concern cases in which a genetic alteration or mutation has led to failure properly to form a particular enzyme; such cases are sometimes easy to detect, since the lack of the enzyme activity amounts to a functional deficiency for which tests can be made.

Much of the most significant work of this kind has been done with the bread mould, *Neurospora crassa*; the American scientists Beadle and Tatum in 1958 received a Nobel prize (together with Lederberg) for their contributions to it. Many mutants of *Neurospora* can be obtained, each of which lacks a different single enzyme. Suppose one such enzyme catalyses the reaction $B \rightarrow C$, which forms part of a metabolic reaction sequence $A \rightarrow B \rightarrow C \rightarrow D$, the product D being a vital cell constituent. The expected consequences of the absence of active enzyme are that the mutant can no longer grow as the original wild type did, unless it is supplied with ready-made D; that C can replace D in making growth possible, but neither A nor B can; and that B accumulates, because of the "genetic block" in the metabolic pathway between B and C. A number of such instances have been worked out. In some cases, they have proved helpful in identifying as yet unknown intermediates in the reaction sequences; broad hints can be got by finding out what accumulates, and which compounds make growth possible.

The major significance of the work, however, lies in the striking demonstration that the enzymes, and hence the reactions of metabolism are genetically determined. Genetic analysis showed that the loss of a single enzyme is a character controlled by a single Mendelian gene. "One gene, one enzyme" was the hypothesis proposed by Beadle and Tatum. (In general, it still seems valid to-day, as long as the meanings of the words are made flexible enough to accommodate the results of finer genetic analysis.) If the gene is altered by mutation—and that seems to mean a change in the chunk of DNA which acts as that gene, by adding or deleting nucleotides, or substituting one for another—then the corresponding enzyme protein is not properly formed, and the course of metabolism is altered in consequence.

Of more direct interest than the work with *Neurospora* are the cases of similar abnormalities in man. Several different kinds of such "inborn errors of metabolism" are known. One of the best characterized is the rare congenital disease called galactosaemia, in which the sugar galactose cannot be normally metabolized. Milk is the main dietary source of galactose, since milk sugar or lactose consists of a residue of galactose joined to one of glucose, and for infants the consequences may be serious, including weight loss, enlargement of the liver, cataract of the eye and even death.

Normally, galactose is handled by phosphorylating it at the expense of ATP, rather as in the case of the very similar sugar glucose (reaction 1 of Fig. 11), except that the phosphate becomes attached to position 1 instead of 6. Galactose-1-phosphate then undergoes an exchange reaction with a compound of glucose and uridine diphosphate (UDP), leaving glucose-1-phosphate and the UDP derivative of galactose.

Galactose + ATP = Galactose-1-phosphate + ADP

Galactose-1-phosphate + UDP-glucose =
$\quad\quad\quad\quad\quad\quad\quad\quad$ Glucose-1-phosphate + UDP-galactose

It is the enzyme for the second step that is missing in galactosaemic individuals. As a result of the genetic block, galactose and its phosphate accumulate, despite the excretion of some galactose in the urine. It seems to be some toxic effect of the abnormally high concentration of galactose, its phosphate or some other derivative that leads to the observed symptoms, because the disease can be completely controlled by excluding galactose from the diet. This means that no milk or milk products may be given; babies are fed on eggs, sugar, margarine and cereal.

Obviously, such dietary treatment cannot be said to cure the disease—it only controls the symptoms. The mutant gene will still be transferred to any children the galactosaemic individual may have. Since the gene behaves as a Mendelian recessive, however, the chances of any offspring suffering from the disease are in general quite small.

The chicken and the egg—what is the basic minimum of life?

It is of interest now to return to a question which was brushed aside earlier (p. 17). The question is, which class of compounds is the most basic to life? It was dismissed previously by saying that all the four major groups—nucleic acids, proteins, carbohydrates and fats—are indispensable for what we at present recognize as life. The emphasis that has since been put on proteins and nucleic acids does not really alter this; without all four groups, living organisms as we know them are inconceivable. Nevertheless, the roles of proteins and nucleic acids have now been seen to be so fundamental that one is tempted to consider them as candidates for the distinction of being regarded as the basic minimum of life.

In favour of the nucleic acids can be urged the fact that virus nucleic acid, alone and unaided, can infect a host cell and multiply there. On the other hand, it cannot reproduce without the aid of the metabolic machinery of a cell, and of this proteins, in their capacities as enzymes, are the major (though not, of course, the only) components. Both a plan and manufacturing facilities are necessary to build living organisms. Trying to allocate priority between proteins and nucleic acids is, in a way, like trying to answer the old question—which came first, the chicken or the egg? The comparison is one which can be followed somewhat further. Taking the chicken to represent growth and the egg to represent reproduction, the one corresponds to the proteins, the other to the nucleic acids. In its "chicken" aspect of vegetative growth, life involves gathering materials and energy, and using them to make more living material; this is basically the responsibility of proteins. In its "egg" aspect of reproduction, life involves making new individuals of the kind; ensuring the substantial accuracy of this is a function of nucleic acids.

A similar impasse is reached if the comparison to the chicken-and-egg problem is taken in a somewhat more literal sense, to refer to priority in time rather than present importance. This raises the ever-fascinating (though hardly pressing) problem of the origin of life. No primordial molecule of either protein or nucleic acid can, as far as can be seen, provide an answer to this vexed question. Neither protein nor nucleic acid alone is known to form a system capable of reproducing itself—each requires the other. Even when both are present, for that matter, as in viruses, the system has a very low degree of self-sufficiency, being incapable of reproduction without the metabolic facilities of a host cell. For this reason, the formation of virus-like particles in the absence of living cells could hardly be counted as the origin of life, and many biologists, indeed, are reluctant to consider viruses to be "alive". Those who care to speculate how life might have arisen have to assume that most of the basic types of molecules were present before biological processes could swing into action. This is a large assumption; the minimum requirement for life remains far greater than it is at all easy to imagine being formed non-biologically. Naturally, one would not expect the origin of life to have occurred by processes that are likely to happen frequently—it is precisely its great improbability that gives the event its momentous significance.

Specific catalysis was earlier (p. 55) called "life's basic device". It is a function, of course, of proteins in their capacities as enzymes. The basic device by which nucleic acids operate, in directing the synthesis of new nucleic acid or protein, seems to consist of determining form by acting as a mould or template. Does this rival specific catalysis as the most basic device of life? Maybe—but it is worth bearing in mind that the two devices are not totally different. In its specificity aspect, enzyme action must also depend on spatial fit; enzyme surface and substrate are presumably related like lock and key (p. 138)—with due allowances in addition, of course, for electrostatic forces and hydrogen or other bonding.

All this emphasizes again the preoccupation of biochemists, like other kinds of biologists, with form and its relation to function (p. 16). The form that primarily interests biochemists differs from that with which other biologists concern themselves in being on the molecular scale of magnitude. Hence the general problem that arises —to relate the molecular level to higher levels of magnitude and organization (p. 97). The role of nucleic acids in shaping development according to heredity is perhaps the biggest of all the particular problems of this kind. Even when it is given that nucleic acid form acts in the first instance by determining protein form, many steps still remain, which are at present difficult even to conceive, before the final expression is reached as a form that we recognize as characteristic of a buttercup, or a frog, or a man.

Index

acetyl 80–86
actin 157–163
adenine 125, 167, 168, 171–173, 176
adenosine 176
adenosine triphosphate, see ATP
alcohol 20, 43, 55–71, 105
amino-acids 18–19, 24, 30–33, 45, 123, 140
ammonium sulphate 20–21, 43, 48–50
AMP 155
antibiotics 137, 141–142
antibodies 23, 48
antimetabolites 139–140
arsenicals 132–135
ascorbate 115, 118
atoxyl 133
ATP 61–65, 77, 106, 143–163, 176
bacteriophage 178
base pairing 172–174
Beadle, G. W., 179
beri-beri 113–115, 119–122
Berzelius, J. J., 56, 68, 70
British Anti-Lewisite 135
Buchner, E., 58–60, 71
buffers 45–49
Calvin, M., 104
carbon monoxide 78
casein 40, 117
cell sap 75, 159, 177
chemotherapeutic index 131
chromatography 24–30, 32, 33, 41–44, 50, 104, 105, 169
chromosomes 165, 170
citrate cycle 80–86
CoI see NAD
CoII see NADP
coenzyme A 80, 82–84
coenzymes 65, 97, 112–128, 159
cooling 74
Cori cycle 155
Crick, F., 171–174
cyanide 92, 101
cytochromes 78–79, 92, 94–95, 98, 100–101, 123
cytoplasm 72
cytosine 168, 171–173
dehydrogenases 79, 84–86, 119, 138
denaturation 20, 21–23, 52, 118
deoxyribose 167, 173
Descartes, R., 88–89
dialysis 20, 50, 118, 119
DNA 166–182
DNP 31–33

Domagk, G., 136
DPN see NAD
drug resistance (acquired) 142
dyes 94, 95, 136, 137
Ehrlich, P., 131–134
Eijkmann, C., 115, 117
electric charges 36–50, 182
electron microscopy 72, 109, 158, 161–163
electrophoresis 44–50
elements 56, 130
enzymes 23, 32, 51–54, 55–71, 83, 93–4, 96, 98, 104, 112, 118, 119, 146, 159, 178–182
Escherichia coli 174, 177
fibrinogen 21, 48, 51–54, 160
fireflies 157
fixing 99
flavoproteins 94–95, 101
folate 118, 123, 139
formaldehyde 104
free energy 144–153
galactosaemia 180
genes 165, 178–180
glucose 61–62, 64, 67, 83, 106, 147–148, 153–155
glycerol 66
glycogen 67, 106, 154–155, 159
guanine 168, 171–173
haemoglobin 78
hair 35
Harden, Sir A., 60, 118
high energy bonds 144–146, 149–150, 152, 155
homogenizer 73
Hopkins, Sir F. G., 116–118
hormones 23, 31
hydrogen bonds 22, 171–174, 182
inborn errors of metabolism 180
insulin 31–35
ion exchange 30, 41–44
iron 78, 91–92
iso-electric point 37, 40, 44, 48
isotopes 102–105, 154, 174
keratin 35
Krebs, Sir H. A., 82–83
lactate 67, 80, 154–155, 159
Lavoisier, A. L. de, 56, 91, 97
Lewisite 135
Liebig, J. von, 57–60
light 106, 157
liver 154–155
lock-and-key analogy 138, 140, 182

INDEX

lysosomes 75–76
Malebranche, N. de 164
Malpighi, M. 164
mechanism 89, 107–111
microscopy 15, 72–73, 99, 109, 160–163, 165
microsomes 75
milk 40, 116–117, 180
mitochondria 72–76, 87, 98, 151, 159
models 87–92, 98, 104, 108, 109
molecular biology 16
muscle 61, 66–68, 97, 154–155, 157–163
myofibrils 159–163
myosin 157–163
NAD 65, 77–80, 86, 94–95, 101, 106, 119, 120, 122–127, 151
NADP 77, 86, 94, 106, 126, 151
nerve 156–157
Neurospora crassa 179
niacin 125
nicotinamide 65, 125–126
nicotinate 115, 118, 122–127, 140
nicotine 122
Nobel prizes 26, 31, 35, 47, 50, 60, 83, 104, 117, 126, 132, 136, 141, 155, 174, 179
nuclei (atomic) 102
nuclei (cell) 72–75, 165, 170, 176
nucleic acids 97, 166–182
nucleotides 125, 168–169, 175, 180
organic chemistry 69, 90, 98
origin of life 181
ovalbumin 21, 48
PAB 137–139
palladium 91
Pasteur, L., 57–60, 70, 114
Pekelharing 117
pellagra 115, 126
penicillin 141–142
peroxides 91
phosphoglycerate 61, 105, 106
photosynthesis 103–106
platinum 91
pneumococci 170
porphyrin 78
preformation 164–166
prontosil 136
prosthetic groups 78, 123–124
protein synthesis 19, 75, 175–182
proteins 17–54, 78, 112, 113, 116, 118, 122, 123, 126, 132, 157, 171, 174, 178, 181
protoplasm 17, 70, 99, 109
purines 167

pyrimidines 167
pyruvate 61, 67, 80, 82, 86, 120–121
radioactivity 102, 105
respiratory chains 77–79, 81, 83, 94–95, 100, 103, 119, 150–152, 154
riboflavin 86, 94, 118, 123, 151
ribose 167
ribosomes 72, 75, 176–177
ribulose diphosphate 105–106
rickets 115
RNA 167, 176–178
salvarsan 133–135
Sanger, F., 31–33
sarcolemma 157, 159
scurvy 113–115
selective toxicity 131, 140
smallpox 131
sodium pump 156–157
spectrophotometer 30, 32
spectroscopy 99–102
spirochaetes 132–135
squid 156
staining 99
starch 106, 117
streptococci 136–138
streptomycin 141–142
succinate 84–86, 138, 151
sulphanilamide 136–138
sulphonamides 136–141
sulphydryl 19, 36, 80, 134–135
supramolecules 124
Swammerdam, J., 165
syphilis 132–134
Takaki, Adml., 114
Tatum, E. L., 179
tetracyclines 142
thiamine 115, 117–124, 128
thrombin 51–54
thymine 168, 171–173, 176
TPN see NADP
TPP 119–124
transformation 170
trypanosomes 132–136
tuberculosis 140–141
ultracentrifugation 50–51, 75, 174
uracil 176–177
viruses 177–178, 181
vitalism 68–70, 89, 107–111
vitamins 97, 112–128, 140
Warburg, O. H., 91–92, 94, 126
Watson, J., 171–174
Wöhler, F., 57, 69
X-ray diffraction 33, 171
yeast 56–71
zymase 59, 118